Science for Sustainable Societies

Scope of the Series

This series aims to provide timely coverage of results of research conducted in accordance with the principles of sustainability science to address impediments to achieving sustainable societies – that is, societies that are low carbon emitters, that live in harmony with nature, and that promote the recycling and re-use of natural resources. Books in the series also address innovative means of advancing sustainability science itself in the development of both research and education models.

The overall goal of the series is to contribute to the development of sustainability science and to its promotion at research institutions worldwide, with a view to furthering knowledge and overcoming the limitations of traditional discipline-based research to address complex problems that afflict humanity and now seem intractable.

Books published in this series will be solicited from scholars working across academic disciplines to address challenges to sustainable development in all areas of human endeavors.

This is an official book series of the Integrated Research System for Sustainability Science (IR3S) of the University of Tokyo.

More information about this series at http://www.springer.com/series/11884

Osamu Saito • Suneetha M Subramanian •
Shizuka Hashimoto • Kazuhiko Takeuchi
Editors

Managing Socio-ecological Production Landscapes and Seascapes for Sustainable Communities in Asia

Mapping and Navigating Stakeholders, Policy and Action

Editors
Osamu Saito
United Nations University
Institute for the Advanced Study
of Sustainability (UNU-IAS)
Shibuya, Tokyo, Japan

Institute for Global Environmental
Strategies (IGES)
Hayama, Kanagawa, Japan

Institute for Future Initiatives (IFI)
The University of Tokyo
Bunkyo, Tokyo, Japan

Shizuka Hashimoto
Graduate School of Agriculture and Life
Sciences
The University of Tokyo
Bunkyo, Tokyo, Japan

Suneetha M Subramanian
United Nations University
International Institute for Global Health
(UNU-IIGH)
Cheras, Kuala Lumpur, Malaysia

Kazuhiko Takeuchi
Institute for Global Environmental
Strategies (IGES)
Hayama, Kanagawa, Japan

Institute for Future Initiatives (IFI)
The University of Tokyo
Bunkyo, Tokyo, Japan

ISSN 2197-7348 ISSN 2197-7356 (electronic)
Science for Sustainable Societies
ISBN 978-981-15-1132-5 ISBN 978-981-15-1133-2 (eBook)
https://doi.org/10.1007/978-981-15-1133-2

This book is an open access publication.

Preface

Core research agendas for sustainability science include the following: (1) co-designing future scenarios and visions with a participatory approach, (2) integrating indigenous and local knowledge (ILK) systems into both scientific knowledge and future scenarios, and (3) the formulation of actions to transform society toward a more sustainable future (Miller et al. 2014; Schneider and Rist 2014; Kishita et al. 2016; Saito 2017).

In 2016, the Intergovernmental Platform on Biodiversity and Ecosystem Services (IPBES) approved a methodological assessment report on scenarios and models of biodiversity and ecosystem services. This report guides experts, stakeholders, and policy makers regarding the use of scenarios and models to perform assessments within IPBES. In this assessment report, "Scenarios" are defined as representations of possible futures for one or more components of a system. In this case, this is achieved with particular emphasis on drivers of change in nature and natural resources, including alternative policy or management options.

While IPBES has identified the development of scenarios as a key to aid decision-makers in identifying potential impacts of different policy options, it currently lacks studies on substantial long-term-scenario approaches (Kok et al. 2017). IPBES emphasizes the importance of ILK together with the social–ecological dynamics of biodiversity and ecosystem services; therefore, engaging with the substantial diversity of local contexts through participatory processes is essential.

To meet this challenge, the authors launched a new project in 2016 named "Predicting and Assessing Natural Capital and Ecosystem Service (PANCES)" which has been funded by Japan's Ministry of the Environment. The aim of this project is to develop an integrated assessment model of social–ecological systems to predict and assess natural and socio-economic values of natural capital and ecosystem services in Japan under various future scenarios (including differing socio-economic conditions and policy options) (PANCES website: http://pances.net/top/). PANCES also promotes multilevel governance of natural capital to maintain and improve "inclusive well-being" and to demonstrate the integrated assessment model at both national and local scales in Japan and beyond.

Responding to the call for papers for "Future scenarios for socio-ecological production landscape and seascape" as a special feature of Sustainability Science in 2017 (Takeuchi et al. 2017), nearly 30 abstracts were submitted. In January 2019, 16 articles were published as a part of this special feature (Saito et al. 2019). In parallel with this special feature, we started editing this book by selecting studies which fit well with the aim and scope of this book from the pool of collected abstracts. Out of 9 chapters, 5 chapters (Chaps. 1–3, 8, and 9) emerged from the PANCES project, 2 chapters (Chaps. 6 and 7) from the International Partnership for the Satoyama Initiative (IPSI: https://satoyama-initiative.org/), and 2 chapters (Chaps. 4 and 5) from external partners in Japan and Bangladesh. The manuscripts outside of the PANCES project were selected on the basis of the quality of the studies, and their practical implications to sustainable management of socio-ecological production landscape and seascape. We would like to acknowledge that most of case studies and review works in this book were funded by the Environment Research and Technology Development Fund (S-15 "Predicting and Assessing Natural Capital and Ecosystem Services" (PANCES), Ministry of the Environment, Japan.

References

IPBES (2016) Summary for policymakers of the methodological assessment of scenarios and models of biodiversity and ecosystem services of the Intergovernmental Science-Policy Platform on Biodiversity and Ecosystem Services. http://www.ipbes.net/publication/methodological-assessment-scenarios-and-modelsbiodiversity-and-ecosystem-services

Kishita Y, Hara K, Uwasu M, Umeda Y (2016) Research needs and challenges faced in supporting scenario design in sustainability science: a literature review. Sustain Sci 11(2):331–347

Kok MTJ, Kok K, Peterson GD, Hill R, Agard J, Carpenter SR (2017) Biodiversity and ecosystem services require IPBES to take novel approach to scenarios. Sustain Sci 12(1):177–181

Miller TR, Wiek A, Sarewitz D, Robinson J, Olsson L, Kriebel D, Loorbach D (2014) The future of sustainability science: a solutions-oriented research agenda. Sustain Sci 9(2):239–246

Saito O (2017) Future science-policy agendas and partnerships for building a sustainable society in harmony with nature. Sustain Sci 12: 895–899. https://doi.org/10.1007/s11625-017-0475-8

Saito O, Hashimoto S, Managi S, Aiba M, Yamakita T, DasGupta R, Takeuchi K (2019) Future scenarios for socio-ecological production landscape and seascape. Sustain Sci 14:1–4. https://doi.org/10.1007/s11625-018-0651-5

Schneider F, Rist S (2014) Envisioning sustainable water futures in a transdisciplinary learning process: combining normative, explorative, and participatory scenario approaches. Sustain Sci 9(4):463–481

Takeuchi K, Saito O, Hashimoto S, Managi S, Aiba M, Yamakita T (2017) Call for papers for "Future scenarios for socio-ecological production landscape and seascape". Sustain Sci 12:633–634. https://doi.org/10.1007/s11625-017-0458-9

Tokyo, Japan   Osamu Saito
Kuala Lumpur, Malaysia   Suneetha M Subramanian
Tokyo, Japan   Shizuka Hashimoto
Tokyo, Japan   Kazuhiko Takeuchi
31 August 2019

Contents

1 Introduction: Socio-ecological Production
  Landscapes and Seascapes............................... 1
  Osamu Saito, Suneetha M Subramanian,
  Shizuka Hashimoto, and Kazuhiko Takeuchi

2 Mapping the Policy Interventions on Marine
  Social-Ecological Systems: Case Study of Sekisei Lagoon,
  Southwest Japan 11
  Mitsutaku Makino, Masakazu Hori, Atsushi Nanami,
  Juri Hori, and Hidetomo Tajima

3 How to Engage Tourists in Invasive Carp
  Removal: Application of a Discrete Choice Model............... 31
  Kota Mameno, Takahiro Kubo, Yasushi Shoji,
  and Takahiro Tsuge

4 The Use of Backcasting to Promote Urban Transformation
  to Sustainability: The Case of Toyama City, Japan.............. 45
  Kazumasu Aoki, Yusuke Kishita, Hidenori Nakamura,
  and Takuma Masuda

5 Traditional Knowledge, Institutions and Human Sociality
  in Sustainable Use and Conservation of Biodiversity
  of the Sundarbans of Bangladesh 67
  Rashed Al Mahmud Titumir, Tanjila Afrin,
  and Mohammad Saeed Islam

6   Lessons Learned from Application of the "Indicators
    of Resilience in Socio-ecological Production Landscapes
    and Seascapes (SEPLS)" Under the Satoyama Initiative   93
    William Dunbar, Suneetha M Subramanian, Ikuko Matsumoto,
    Yoji Natori, Devon Dublin, Nadia Bergamini, Dunja Mijatovic,
    Alejandro González Álvarez, Evonne Yiu, Kaoru Ichikawa,
    Yasuyuki Morimoto, Michael Halewood, Patrick Maundu,
    Diana Salvemini, Tamara Tschenscher, and Gregory Mock

7   Place-Based Solutions for Conservation and Restoration
    of Social-Ecological Production Landscapes
    and Seascapes in Asia   117
    Raffaela Kozar, Elson Galang, Jyoti Sedhain, Alvie Alip,
    Suneetha M Subramanian, and Osamu Saito

8   Mapping the Current Understanding of Biodiversity
    Science–Policy Interfaces   147
    Ikuko Matsumoto, Yasuo Takahashi, André Mader,
    Brian Johnson, Federico Lopez-Casero, Masayuki Kawai,
    Kazuo Matsushita, and Sana Okayasu

9   Synthesis: Managing Socio-ecological Production
    Landscapes and Seascapes for Sustainable Communities
    in Asia   171
    Osamu Saito, Suneetha M Subramanian,
    Shizuka Hashimoto, and Kazuhiko Takeuchi

Chapter 1
Introduction: Socio-ecological Production Landscapes and Seascapes

Osamu Saito, Suneetha M Subramanian, Shizuka Hashimoto, and Kazuhiko Takeuchi

Abstract This book presents up-to-date analyses of community-based approaches to the sustainable resource management of socio-ecological production landscapes and seascapes (SEPLS) in areas where a harmonious relationship between the natural environment and the people who inhabit it is essential to ensure community and environmental well-being as well as to build resilience in the ecosystems that support this well-being. This chapter introduces the key concepts and approaches, objectives, and organization of this book.

Keywords Socio-ecological production landscapes · Indigenous and local knowledge · Science–policy interface · Ecosystem services · Visualization · Mapping · Stakeholder analysis

O. Saito (✉)
United Nations University Institute for the Advanced Study of Sustainability (UNU-IAS), Shibuya, Tokyo, Japan

Institute for Global Environmental Strategies (IGES), Hayama, Kanagawa, Japan

Institute for Future Initiatives (IFI), The University of Tokyo, Bunkyo, Tokyo, Japan
e-mail: saito@unu.edu

S. M. Subramanian
United Nations University International Institute for Global Health (UNU-IIGH), Cheras, Kuala Lumpur, Malaysia

S. Hashimoto
Graduate School of Agriculture and Life Sciences, The University of Tokyo, Bunkyo, Tokyo, Japan

K. Takeuchi
Institute for Global Environmental Strategies (IGES), Hayama, Kanagawa, Japan

Institute for Future Initiatives (IFI), The University of Tokyo, Bunkyo, Tokyo, Japan

© The Author(s) 2020
O. Saito et al. (eds.), *Managing Socio-ecological Production Landscapes and Seascapes for Sustainable Communities in Asia*, Science for Sustainable Societies, https://doi.org/10.1007/978-981-15-1133-2_1

1.1 Socio-ecological Production Landscapes and Seascapes

A landscape can be defined as "an area, as perceived by people, whose character is the result of the action and interaction of natural and/or human factors" (Council of Europe 2000). Socio-ecological production landscapes and seascapes (SEPLS) can be characterized by a mosaic of different ecosystem types: secondary forests, timber plantations, farmlands, irrigation ponds, wetlands, grasslands, beaches, and coastal zones, as well as human settlements. SEPLS are managed via interactions between ecosystems and humans to create various ecosystem services for human well-being (Japan Satoyama Satoumi Assessment (JSSA) 2010; Takeuchi 2010; Duraiappah et al. 2012). In Japan, the term satoyama is used for such landscapes (Fig. 1.1 top), while satoumi refers to such seascapes (Fig. 1.1 bottom). The term "cultural landscapes" is often used synonymously for similar landscapes where people have developed and sustainably managed the landscape over a long period of time. According to UNESCO (2008), cultural landscapes represent the "combined works of nature and of man" and are illustrative of the evolution of human societies and settlements over time. Examples of cultural landscapes include the rice terrace landscapes of the Philippines, the Black Forest mountain range in southern Germany, extensively used mountain grasslands in the European Alps, and the dehesa agroforestry landscapes on the Iberian Peninsula (Plieninger and Bieling 2012).

According to the Millennium Ecosystem Assessment (2005), ecosystem services are defined as benefits obtained from ecosystems, including provisioning services such as food and water; regulating services such as the regulation of floods, droughts, and diseases; supporting services such as soil formation and nutrient cycling; and cultural services such as recreational, spiritual, and other nonmaterial benefits. The Intergovernmental Science-Policy Platform on Biodiversity and Ecosystem Services (IPBES) extended the concept of ecosystem services to nature's contributions to people (NCPs) which can capture all types of contributions by nature, whether these contributions result in gains or losses for humans. The notion of NCP also recognizes the central and pervasive role that culture plays in defining all links between people and nature and in emphasizing and operationalizing the role of indigenous and local knowledge in understanding NCPs (Fig. 1.2). IPBES identified 18 such categories for reporting NCPs within a generalized perspective organized into three partially overlapping groups: regulating, material, and nonmaterial NCPs. Even though the concept of NCP is formally approved by IPBES and used in IPBES assessments, NCP is still quite a new concept and requires a transitional period to be widely acknowledged. Therefore, this book uses the term ecosystem services to represent various tangible and intangible values provided by nature.

The categories in gray are part of the framework but not the focus of Díaz et al. (2018). Concepts pointed to by the arrowheads replace or include concepts near the arrow tails. Concepts in dotted-line boxes are no longer used; following the present view of the Millennium Ecosystem Assessment community, supporting ecosystem services are now components of nature or (to a lesser extent) regulating NCPs.

Fig. 1.1 Illustrations of satoyama (top) and satoumi (bottom) (Saito and Shibata 2012)

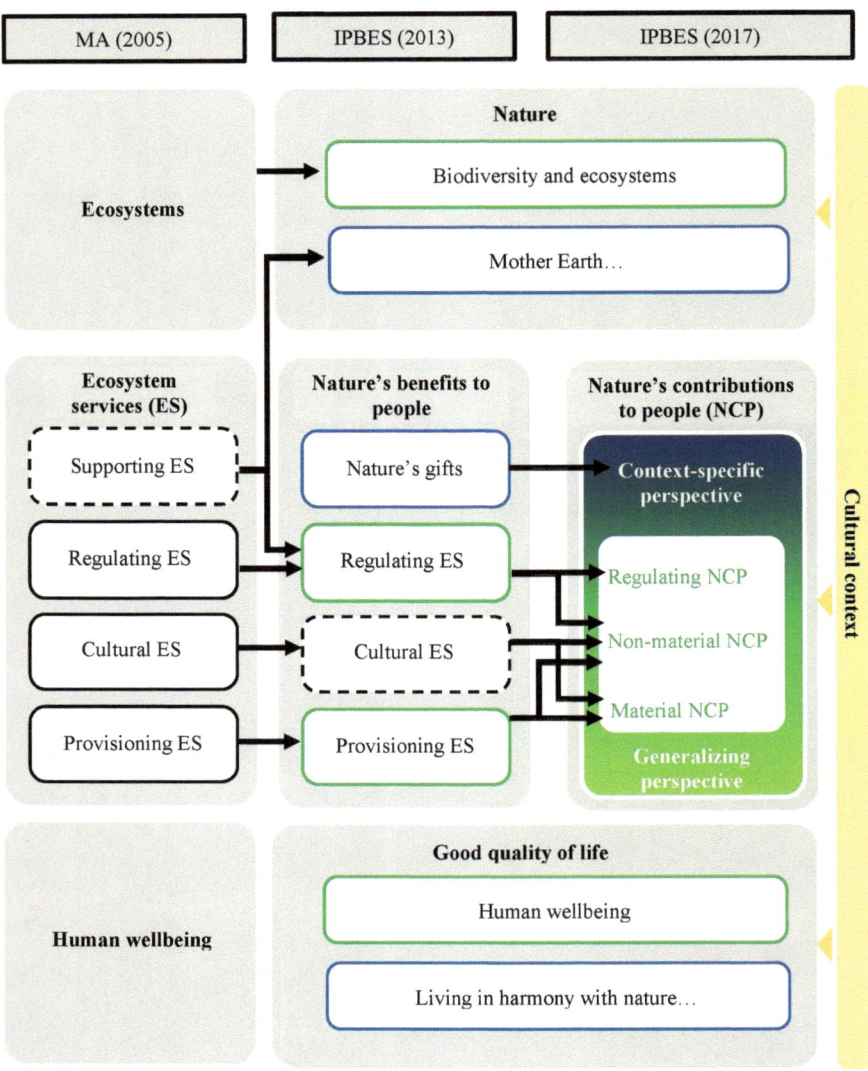

Fig. 1.2 Evolution of nature's contributions to people (NCPs) and other major categories in the IPBES conceptual framework (1) with respect to the concepts of ecosystem services and human well-being as defined in the Millennium Ecosystem Assessment (2) (Modified from Díaz et al. 2018)

1.2 Challenges and Opportunities for Contemporary SEPLS

1.2.1 Challenges

According to the IPBES Global Assessment (IPBES 2019), the rate of global change in nature during the past 50 years is unprecedented in human history even though the rate of change differs between regions and countries. The direct drivers of change in nature with the largest global impact have been changes in land and sea use, the direct exploitation of organisms, climate change, pollution, and invasions of alien species. These five direct drivers are in turn underpinned by societal values and behaviors that include production and consumption patterns, human population dynamics and trends, trade, technological innovations, and local to global governance.

The Asia–Pacific region is home to nearly 60% (4.5 billion) of the current global population, 52% (400 million) of the 767 million global poor, and as much as 75% of the global population of 370 million indigenous people. Most of the latter have distinct but increasingly threatened traditions and cultures and have been maintaining their livelihoods in harmony with nature and managing landscapes and seascapes for generations (IPBES 2018). In addition to rapid economic growth, globalization, urbanization, infrastructure development, unsustainable use, and invasive alien species, the IPBES Asia–Pacific Regional Assessment (2018) highlighted a decline in traditional agrobiodiversity, along with its associated indigenous and local knowledge, due to a shift toward the intensification of agriculture with a small number of improved crop species and varieties.

The Japan Biodiversity Outlook (Japan Biodiversity Outlook Science Committee 2010) and Japan's National Biodiversity Strategy and Action Plan (Ministry of the Environment, Japan 2012) have recognized four biodiversity crises that have been faced by Japan in recent years. The first crisis stems from development, overexploitation, and water contamination. This crisis has been particularly influential; however, the situation has been mitigated by the regulation of developmental activities and the slowing of economical development. The second crisis is caused by the reduced use and insufficient management of SEPLS. This tendency continues to intensify due to depopulation and the aging of populations in rural areas in Japan. Factors contributing to the third crisis include invasive alien species and chemical substances introduced by humans. Climate change, as the fourth crisis, has reinforced the effects of the other crises, causing serious concern regarding certain particularly vulnerable ecosystems.

1.2.2 Opportunities

IPBES Global Assessment (IPBES 2019) stressed that "goals for conserving and sustainably using nature and achieving sustainability cannot be met by current trajectories, and goals for 2030 and beyond may only be achieved through transformative changes across economic, social, political, and technological factors." It is necessary for us to conserve, restore, and use nature sustainably while simultaneously meeting other global societal goals via extensive efforts to foster transformative change. Transformations toward sustainability can be triggered by the following key leverage points: (1) visions of a good life; (2) total consumption and waste; (3) values and actions; (4) inequalities; (5) justice and inclusion in conservation; (6) externalities and telecoupling; (7) technology, innovation, and investment; and (8) education and knowledge generation and sharing (IPBES 2019). The transformation pathways will vary depending on the context, with different challenges and needs in developing and developed countries. Therefore, "risks related to inevitable uncertainties and complexities in transformations toward sustainability can be reduced through governance approaches that are integrative, inclusive, informed, and adaptive" (IPBES 2019).

In the Asia–Pacific region, regional cooperation for the transboundary conservation of threatened landscapes and seascapes is expanding and showing positive outcomes (IPBES 2018). Biodiversity-rich and threatened terrestrial, marine, and wetland ecosystems transcend political boundaries. Transboundary conservation initiatives take different forms including upstream–downstream river basins initiatives (e.g., in the Mekong Delta Basin), ridge-to-reef arrangements (coral reef conservation and management through community-based approaches emphasizing land–sea connectivity), and regional cooperative agreements (IPBES 2018).

As one such transboundary/international conservation initiative, the Japanese Government and United Nations University launched a new international initiative called "the Satoyama Initiative," which aims to promote sustainable production landscapes and seascapes via a broader global recognition of their value (Takeuchi 2010). This initiative promotes developing an international network of organizations working on SEPLS to share knowledge and best practices on a global scale to alleviate some of the problems caused by the loss of biodiversity. Globally Important Agricultural Heritage Systems, coordinated by the Food and Agriculture Organization, is another international initiative that promotes public understanding, awareness, and the national and international recognition of agricultural heritage systems including SEPLS.

1.3 Sustainability Science Research and SEPLS

Understanding SEPLS and the forces of change that can weaken their resilience requires the integration of knowledge across a wide range of academic disciplines as well as from indigenous knowledge and experience. Moreover, given the wide variation in the socio-ecological makeup of SEPLS globally, as well as in their political and economic contexts, individual communities will be at the forefront of developing appropriate measures for their unique circumstances. This in turn requires robust communication systems and broad participatory approaches.

Sustainability Science (SuS) has emerged as a new transdisciplinary academic discipline in the last decade and offers a new, broad-perspective approach to deal with complex, long-term global issues, such as human-induced climate and ecosystem changes. It aims to promote solutions that contribute to rebuilding a sound relationship between human societies and the environment (UNESCO 2017). SuS research is highly integrated, participatory, and solution driven and, as such, is well suited to the study of SEPLS. Using case studies, literature reviews, and SuS analyses, this book explores various approaches to stakeholder participation, policy development, and appropriate actions for the future of SEPLS. It provides communities, researchers, and decision-makers at various levels with new tools and strategies for exploring scenarios and creating future visions for sustainable societies.

This book presents up-to-date experience and analyses of various approaches to the sustainable resource management of SEPLS, primarily based on experiences in Asia.

1.4 Objectives and Organization of the Book

SEPLS are areas in which the majority of inhabitants rely on the well-being of the landscape or seascape ecosystem. By definition, a harmonious relationship between the natural environment and the people who inhabit it is essential to ensure community and environmental well-being as well as to build resilience in the ecosystems that support this well-being. Understanding SEPLS and the forces of change that can weaken their resilience requires the integration of knowledge across a wide range of academic disciplines as well as from indigenous knowledge and experience. Moreover, given the wide variation in the socio-ecological makeup of SEPLS around the world, as well as in their political and economic contexts, individual communities will be at the forefront of tailoring the approaches necessary to their unique circumstances. Including SuS research approaches and integration of indigenous and local knowledge systems and scientific knowledge, this book explores various approaches to stakeholder participation, policy development, and appropriate action for the future of SEPLS. By providing such approaches and tools, this book shows how decision-makers and policy planners can promote robust collaborations between different stakeholders that will contribute to the effective implementation of conservation and development policies for sound resource management in SEPLS.

While Chaps. 2–5 cover specific case studies of land/seascapes in Japan (Chaps. 2–4) and in Bangladesh (Chap. 5), Chaps. 6–8 consist of a series of review articles that explore lessons learned from assessing resilience in SEPLS (Chap. 6), solutions for the sustainable management of SEPLS in Asia (Chap. 7), and the effectiveness of biodiversity science–policy interfaces (SPIs) from local to global scales (Chap. 8). The book highlights various approaches to navigate the sustainable resource management of SEPLS from local to global scales.

Focusing on marine systems, *Chap. 2* examines the interrelationships between sectoral policy interventions by various marine-related ministries and the entire structure of integrated ocean policies. Focusing on the Sekisei Lagoon, Okinawa Prefecture, on the southeastern tip of the Japanese archipelago, this study demonstrates clear structural and functional interlinkages between relevant sectors, further highlighting the close connections between various stakeholders at the ecological level.

Chapter 3 focuses on engaging tourists in addressing the issue of invasive fish species (carp) via a choice experiment survey conducted in Amami Oshima, Japan, to quantify the willingness of tourists to participate in invasive carp removal in nature-based tourism.

Given the rapid urbanization of the Asian region, we also focus on approaches to ensure sustainability in urban contexts. Using an example from the city of Toyama in Japan, *Chap. 4* highlights how urban systems can move toward sustainability using an envisioning method and further identifies pathways to reach such visions. The chapter focuses on participatory approaches in urban contexts and identifies ways of bringing together various perspectives to enable planning.

Chapter 5 highlights how local institutions and traditional knowledge can be incorporated when addressing the sustainable use and conservation of biodiversity, focusing on experiences from the Sunderbans area in Bangladesh.

Chapter 6 dwells on this issue as it narrates experiences from the Satoyama Initiative in the development and use of indicators of resilience in SEPLS in different regions of the world. This indicator toolkit is being used to assess, consider, and monitor the circumstances of a landscape or seascape, identifying important issues and ultimately improving their resilience.

Chapter 7 identifies various categories of solutions for the sustainable management of SEPLS based on the experiences of partners from the South, East, and Southeast Asian countries of the International Partnership for the Satoyama Initiative.

Chapter 8 provides a review of the effectiveness of different biodiversity SPIs, which play a vital role in navigating policies and actions with a sound evidence base. Based on a systematic review of 96 SPI studies from local to global scales, this chapter examines the SPIs in terms of their perceived credibility, relevance, and legitimacy.

Chapter 9 consolidates Chaps. 2–8 to identify key messages and future actions to improve the science–policy–society interface for SEPLS, including future research directions.

Acknowledgements This book was funded by the Environment Research and Technology Development Fund (S-15 "Predicting and Assessing Natural Capital and Ecosystem Services" (PANCES), Ministry of the Environment, Japan).

References

Council of Europe (2000) The European Landscape Convention. Council of Europe, Strasbourg

Díaz S, Pascual U, Stenseke M, Martín-López B, Watson BT, Molnár Z, Hill R, Chan KM, Baste IA, Brauman K, Polasky S, Church A, Lonsdale M, Larigauderie A, Leadley PA, van Oudenhoven APE, van der Plaat F, Schröter M, Lavorel S, Aumeeruddy-Thomas Y, Bukvareva E, Davies K, Demissew S, Erpul G, Failler P, Guerra CA, Hewitt CL, Keune H, Lindley S, Shirayama Y (2018) Assessing nature's contributions to people. Science 359(6373):270–272. https://doi.org/10.1126/science.aap8826

Duraiappah AK, Nakamura K, Takeuchi K, Watanabe M, Nishi M (eds) (2012) Satoyama-satoumi ecosystems and human well-being: socio-ecological production landscapes of Japan. United Nations University Press, Tokyo

IPBES (2018) Summary for policymakers of the regional assessment report on biodiversity and ecosystem services for Asia and the Pacific of the Intergovernmental Science-Policy Platform on Biodiversity and Ecosystem Services. Karki M, Senaratna Sellamuttu S, Okayasu S, Suzuki W, Acosta LA, Alhafedh Y, Anticamara JA, Ausseil AG, Davies K, Gasparatos A, Gundimeda H, Faridah-Hanum I, Kohsaka R, Kumar R, Managi S, Wu N, Rajvanshi A, Rawat GS, Riordan P, Sharma S, Virk A, Wang C, Yahara T, Youn YC (eds) IPBES Secretariat, Bonn. 41p. https://www.ipbes.net/system/tdf/spm_asia-pacific_2018_digital.pdf?file=1&type=node&id=28394

IPBES (2019) Summary for policymakers of the global assessment report on biodiversity and ecosystem services of the Intergovernmental Science-Policy Platform on Biodiversity and Ecosystem Services, IPBES/7/10/Add.1. https://www.ipbes.net/system/tdf/ipbes_7_10_add-1-_advance_0.pdf?file=1&type=node&id=35245

Japan Biodiversity Outlook Science Committee (2010) Japan biodiversity outlook 2. Nature Conservation Bureau, Ministry of the Environment, Japan

Japan Satoyama Satoumi Assessment (JSSA) (2010) Satoyama-satoumi ecosystems and human wellbeing: socio-ecological production landscapes of Japan (summary for decision makers). United Nations University, Tokyo

Millennium Ecosystem Assessment (MA) (2005) Ecosystem and human well-being – summary for decision makers. Island Press, Washington, DC

Ministry of the Environment, Japan (2012) National biodiversity strategy of Japan 2012-2020: roadmap towards the establishment of an enriching society in harmony with nature

Plieninger T, Bieling C (2012) Resilience and the cultural landscape: understanding and managing change in human-shaped environments. Cambridge University Press, Cambridge

Saito O, Shibata H (2012) Satoyama–satoumi and ecosystem services: a conceptual framework. In: Duraiappah AK, Nakamura K, Takeuchi K, Watanabe M, Nishi M (eds) Satoyama-satoumi ecosystems and human well-being: socio-ecological production landscapes of Japan. United Nations University Press, Tokyo, pp 17–59

Takeuchi K (2010) Rebuilding the relationship between people and nature: the Satoyama Initiative. Ecol Res 25:891–897

UNESCO (2008) Operational guidelines form the implementation of the World Heritage Convention. UNESCO World Heritage Centre, Paris

UNESCO (2017) Practical guidelines to apply sustainability science frameworks, JAK/2017/PI/H/6. http://unesdoc.unesco.org/images/0025/002593/259341E.pdf

Chapter 2
Mapping the Policy Interventions on Marine Social-Ecological Systems: Case Study of Sekisei Lagoon, Southwest Japan

Mitsutaku Makino, Masakazu Hori, Atsushi Nanami, Juri Hori, and Hidetomo Tajima

Abstract Using a case of the Sekisei Lagoon, Okinawa Prefecture, the southeastern tip of Japanese archipelago, this chapter discussed the interrelationships among the sectoral policy interventions by various marine-related ministries, and the whole structure of the integrated ocean policy. First, we developed the Social-Ecological Systems (SES) Schematic, which summarized the main ecosystem structures, functions, use types, and the stakeholders relating to the Sekisei Lagoon. Then, sectoral policy interventions by various ministries were overlaid onto the SES schematic to graphically show their interrelationships. We found that the ecosystem structure and functions used by one sector is closely connected to other structures and functions, which are then used by other sectors. In other words, all the stakeholders in the social system are closely interlinked at the ecological system level. Secondly, all in all, sectoral policy interventions by various ministries are covering almost all part of the Sekisei Lagoon SES, and therefore, the total coordination of the sectoral policy interventions and the creation of the synergy effects are required. In this process, the cabinet office and the local government

M. Makino (✉)
Atmosphere and Ocean Research Institute, University of Tokyo, Chiba, Japan
e-mail: mmakino@aori.u-tokyo.ac.jp

M. Hori
National Research Institute of Fisheries and Environment of Inland Sea, Japan Fisheries Research and Education Agency, Yokohama, Japan

A. Nanami
Research Center for Subtropical Fisheries, Seikai National Fisheries Research Institute, Japan Fisheries Research and Education Agency, Ishigaki, Japan

J. Hori
Education Unit for Studies on Connectivity of Hills, Humans and Oceans, Kyoto University, Kyoto, Japan

H. Tajima
Japan Fisheries Education and Research Agency, Yokohama, Japan

© The Author(s) 2020   11
O. Saito et al. (eds.), *Managing Socio-ecological Production Landscapes and Seascapes for Sustainable Communities in Asia*, Science for Sustainable Societies, https://doi.org/10.1007/978-981-15-1133-2_2

will play the important roles. Finally, this SES schematic can be used as a boundary object to facilitate the knowledge exchanges among various stakeholders including the policy makers, practitioners, and researchers, to share the common understandings of the current situation, and to cocreate the policy interventions for the sustainable uses of Sekisei Lagoon.

Keywords Integrated ocean policy · Sectoral policy interventions · Social-ecological systems (SES) schematic · Mapping · Stakeholders · Sekisei Lagoon

2.1 Introduction

2.1.1 Ocean Policy in Japan

In Japan, eight ministries are implementing marine-related policies, i.e., Ministry of Agriculture, Forestry and Fishery: MAFF, Ministry of Environment: MoE, Ministry of Land, Infrastructure, Transport and Tourism: MLIT, Ministry of Education, Culture, Sports, Science and Technology: MEXT, Ministry of Economy Trade and Industry: METI, Ministry of Internal Affairs and Communications: MIC, Ministry of Foreign Affairs: MFA, and Ministry of Defense: MoD. Each ministry has its own policy missions, legal basis, marine use-types under the jurisdiction, and stakeholders. Traditionally, each ministry has been implementing its specific policy interventions separately and independently (often called as "sectoral policy interventions"). As the result, it has not been clear enough how the Japanese ocean policy, as a whole, would deal with the up-to-date issues such as environmental degradation, overfishing, integrated coastal zone management, national security, etc. Therefore, in 2007, in order to promote the coordination among sectoral policies interventions by eight ministries and to formulate the integrated ocean policy, the Basic Act on Ocean Policy (hereafter, the Act) was legislated (Sakamoto 2018; Makino 2011). Based on the Act, the Headquarter for the Ocean Policy was formulated at the Cabinet Secretariat of Japan (moved to the Cabinet Office in 2017), headed directly by the Prime Minister.

2.1.2 Objective of This Chapter

Conventionally, a lot of assessments and analysis have been conducted on the design or effectiveness of the sectoral policy interventions, and therefore analytical methodologies for that purpose have been developed (For example, Dunn 2016; Weimer and Vining 2017). Recently, many studies have been conducted on the interrelationships among the sectoral policy interventions by each ministry, and on the comprehensive design or the effectiveness of the integrated ocean policy as a whole (for example, Cicin-Sain and Knecht 1998; Guneroglu 2015). Also, marine ecosystems have high uncertainties and fluctuations with very limited scientific knowledge. And because the objective of the ecosystem conservation are "the matter of societal choices" (Principle 1 of the Ecosystem Approach of the Convention on Biological Diversity), the expected effects by policy interventions to the stakeholders are important information for the social agreement and the effective co-implementation of the policy interventions (Ban et al. 2013; Ehler 2014; Kittinger et al. 2014; Schultz et al. 2015; Bodin 2017). Therefore, using a case of the Sekisei Lagoon, Ishigaki City, Okinawa Prefecture, Japan (Fig. 2.1), this chapter tried to understand and to analyze the whole structure of the marine-related policy measures introduced by various ministries, and their interrelationships with not only the ecosystem structures and functions but also with the various use types and stakeholders. Especially, based on the first Basic Principle of the Act (Article 2), this chapter focused on the policy interventions for " Harmonization of the Development and Use of the Oceans with the Conservation of Marine Environment".

There are many types of uses in the Sekisei Lagoon. Traditionally, the fisheries resources at the very nearshore coastal area have been utilized by the local people at the daily basis and it constitutes an important part of local culture (Sugimoto 2016). The main commercial uses are the fishery and tourism. Because it is remoted islands area, marine transport (people, food, goods, etc.) are also very important. In addition, the environmental education and research activities by local schools and NGOs (for example, the WWF Japan) are very active here. On the other hand, for the last few decades, the coral reefs have been widely destroyed and deteriorated presumably by the over uses by various stakeholders, impacts from the land, and the effects from climate change. To deal with this issue, the "Sekisei Lagoon Nature Restoration Committee" was established and variety of policy interventions have been implemented here (Lou et al. 2017).

Fig. 2.1 Location of the Sekisei Lagoon, Ishigaki City, Okinawa Prefecture

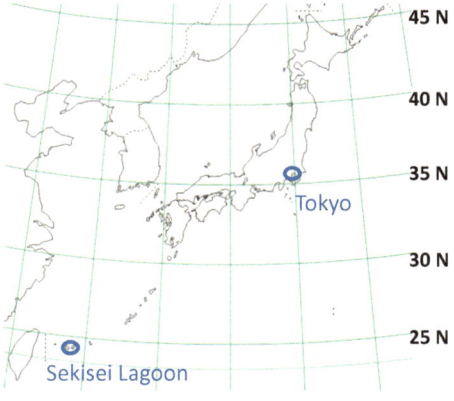

2.2 Method

2.2.1 The Social-Ecological Systems (SES) Schematic

There are many ways to conceptualize the interrelationships between ecological system and social systems (For example, Ostrom 2009, Berkes et al. 2014, Bodin 2017, Diaz et al. 2018, etc.) Some studies drew social-ecological diagram based on the interviews/workshops with local stakeholders. For example, Howard et al. (2013) dealt with the climate change in Australia and discussed the marine biodiversity conservation scenarios with stakeholders. Tiller et al. (2017) discussed about the governance of Norwegian salmon aquaculture with local stakeholders and government officers, and developed the conceptual map. In this chapter, because our study focused on the policy interventions for harmonization of the uses and conservation (Article 2 of the Act), we tried to describe the interrelationship among the main ecosystem structures, ecosystem functions, ecosystem use types, and stakeholders, in the following manner.

Firstly, the ecological scientists in the co-authors conducted the literature reviews and hearing survey to biologists, ecologists, fisheries scientists, etc., who are doing natural science researches on the Sekisei Lagoon, and identified the important ecosystem structure and functions there. On the other hand, the social scientists in the coauthors conducted the web-based survey to identify the main stakeholders relating to the marine ecosystem services (Hori et al. 2017). Based on the result, they conducted the field stakeholder analysis in Ishigaki City (snowball method interviews) to identify the main use-types of the Sekisei Lagoon ecosystem services. They interviewed to the local fishers, agricultural farmers, local/national government officers, coast guard, local researchers, environmental NGOs (including WWF-Japan), tourism association, diving association, restaurants, ferry company, and local hotels. Finally, these results are combined into an integrated diagram, called as the Sekisei Lagoon Social-Ecological Systems (SES) Schematic.

2.2.2 Review of the Policy Interventions

Based on the Sekisei Lagoon SES Schematic, literature reviews and interviews were conducted to the local/central government officers and environmental policy experts, and a list of main legal basis (acts) relating to the policy interventions to the Sekisei Lagoon by various ministries was developed. Then, coauthors identified the components within the Sekisei Lagoon SES Schematic that each policy intervention by each ministry is targeting. Finally, the coverages of the overall policy interventions by various ministries were graphically summarized over the SES Schematic (Makino et al. 2018).

2.3 Results

Figure 2.2 is the Sekisei Lagoon SES Schematic developed by the authors. It shows that the ecosystem functions in the coastal areas, in which the reef building corals locate, are closely linked to the land, intertidal, and the offshore ecosystems. For example, juvenile fish grew up in the nursery ground at the intertidal area, which is under the heavy influences from the land discharges, and then inhabit to the coral reef areas where fisheries operations are conducted. Also, it shows the wide-ranging ecosystem functions have been utilized by various use types and stakeholders. For example, fisheries sector (fishers, processors, retailers, etc.) are relying on the ecosystem functions such as nursery ground, spawning ground, feeding area, etc., for harvesting fish. The same functions are also utilized by other sectors such as the tourism sector.

Table 2.1 summarized the government bodies, legal basis (acts), and their target components in the Sekisei Lagoon SES Schematic. There are more than 50 acts under the charge of at least five ministries and local government (Okinawa Prefecture and Ishigaki City). Note that, the policy measures relating to the MoD, are important, but not directly relating to the ecosystem of the Sekisei Lagoon. Therefore, they are excluded from this analysis. Also, there is a territorial issue in Ishigaki City, and therefore policy interventions by the MFA are important, but it was excluded based on the same reason. METI is mainly in charge of the marine renewable energy development such as offshore wind power, or sea bottom resource development such as submarine manganese nodules or cobalt-rich crust, both of them are not existing around the Sekisei Lagoon at the moment. MIC is mainly in charge of the information and communication technologies for vessels or communities in the remoted islands, which is again not the case in the Sekisei Lagoon.

Finally, Figs. 2.3, 2.4, 2.5, 2.6, 2.7, and 2.8 graphically summarized the coverage of each sectoral policy interventions on the SES Schematic. The dark color shows the direct and stronger relationships (double circles in Table 2.1), while light color shows the indirect and weaker relationships (single circles in Table 2.1). The characteristics of each sectoral policy are clearly illustrated. For example, MAFF measures (Fig. 2.4) are targeting the ecosystem functions and structures directly relating to the fisheries resources and agriculture on land, while MoE measures (Fig. 2.3) are more about the water quality or waste/soil management, as well as the nature restoration activities for coral reefs such as the management of the "Sekisei Lagoon Nature Restoration Committee". MLIT (Fig. 2.5) is in charge of tourism and transport, as well as the land development on the island. As shown in Fig. 2.6, MEXT's main policy target is research, education, monitoring, and the conservation of the natural/cultural monuments. The Cabinet Office (Fig. 2.8) are covering almost all part of the SES based on the national-scale plans or strategies such as Ocean Basic Plan, Marine Biodiversity Conservation Strategy, Okinawa Promotion and Development Plan, etc. Similarly, the local government has similar plans/strategies and covering almost all the SES components (Fig. 2.7).

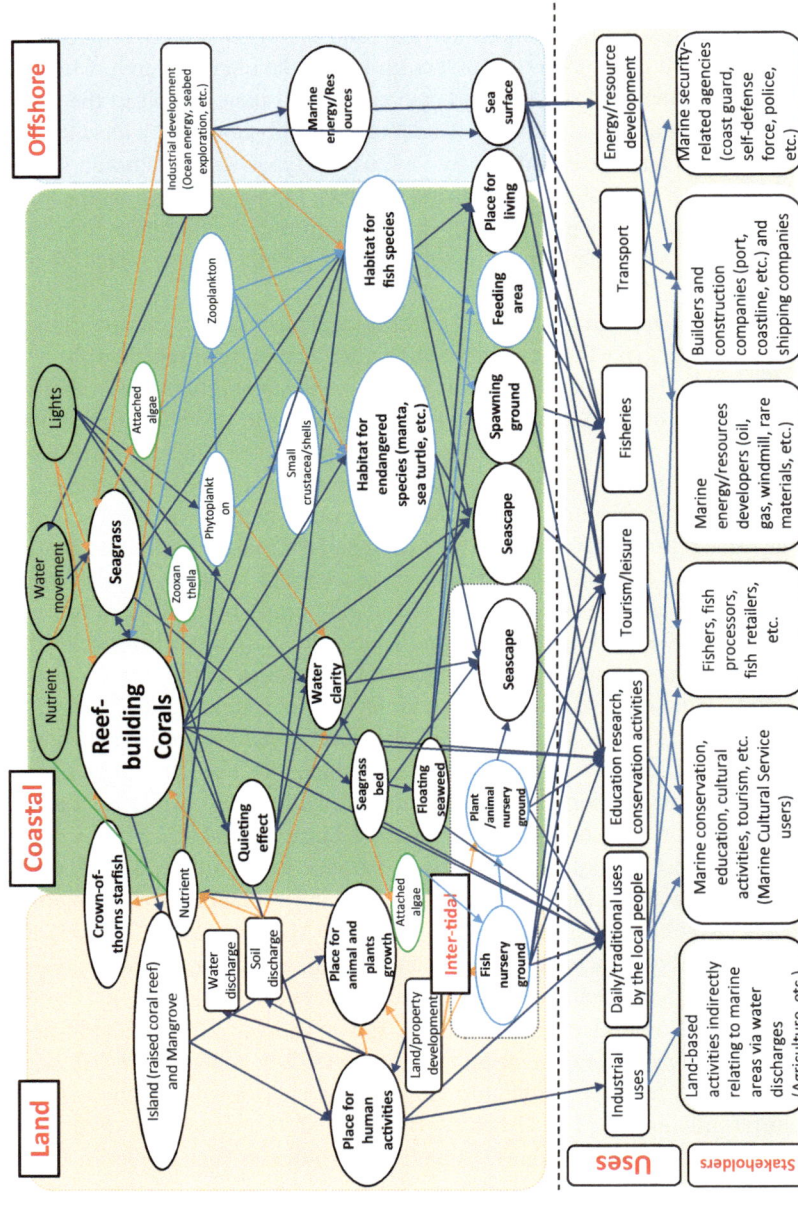

Fig. 2.2 The social-ecological systems (SES) schematic for the Sekisei lagoon. Arrows in Navy blue means ecosystem functions, light blue means food web (prey–predator relationship), and orange means other relations

Table 2.1 Summary of government bodies and legal basis for the policy interventions to the Sekisei Lagoon SES components

Government Body	Legal/Institutional Basis	Land					Inter-tidal									Coastal area						Offshore			Others	Use types						
		Inland formation and Mangrove	Place for human activities	Property/ land development	Water discharge	Soil discharge	Nursery ground for animal and plants	Fish nursery ground	Seascape	Reef-building corals	Seagrass	Nutrient	Cross-of-thru airfall	Quantity g effect	Water clarity	Phys/ Zoo plankton	Small crustacea shells	Seascape	Habitat for endangered species	Habitat for fish species	Spawning and feeding ground	Sea surface	Place for industrial development	Marine energy sources	Water movement	Lights	Industrial development	Daily/ traditional uses by locals	Education, research, conservation activities	Tourism/ leisure	Fisheries	Transport	Energy/ seabed exploration
MAFF	Agricultural Land Act																																
	Act on Establishment of Agricultural Promotion Regions																																
	Land Improvement Act																																
	Forest Act																																
	Coast Act																																
	Fisheries Basic Act																																
	Fishery Act																																
	Act on the Protection of Fishery																																
	Fishery Coordinating Regulation																																
	Marine Resources Development Promotion Act																																
	Fishery Cooperative Act																																
	Multifunctional Roles of Fisheries																																
	Act on Development of Fishing Ports and Grounds																																
	Coastal Fisheries Grounds Enhancement and Development Program Act																																
	Marine Coast Coordinating Council																																
MOE	Natural Parks Act																																
	Basic Environment Act																																
	Basic Act on Biodiversity																																
	Nature Conservation Act																																
	Protection and Control of Wild Birds and Mammals and Hunting Management Law																																
	Act on Conservation of Endangered Species of Wild Fauna and Flora																																
	Environmental Impact Assessment Act																																
	Act on the Promotion of Nature Restoration																																
	Act on Prevention of Marine Pollution and Maritime Disaster																																
	Waste Management and Public Cleaning Act																																
	Water Pollution Prevention Act																																
	Basic Policy on the Comprehensive and Effective Promotion of Measures Against Articles that Drift Ashore																																
	Act on the Promotion of Environmental Conservation Activities through Environmental Education																																
	Ecotourism Promotion Act																																
	Act on Reclamation of Publicly-owned Water Surface																																
MLIT	National Spatial Planning Act																																
	National Land Use Planning Act																																
	City Planning Act																																
	Landscape Act																																
	Tourism-based Country Promotion Basic Act																																
	Coast Act																																
	Act on Prevention of Marine Pollution and Maritime Disaster																																
	Port and Harbor Act																																
MEXT	Social Education Act																																
	Act on Protection of Cultural Properties																																
	Basic Act on Education																																
Cabinet Office	Act on Special Measures for the Promotion and Development of Okinawa																																
	Basic Act on Ocean Policy																																
Okinawa Pref.	the okinawa prefecture land preservation ordinance																																
	the okinawa prefecture basic environment ordinance																																
	the okinawa prefecture red soil runoff prevention ordinance																																
	the okinawa prefecture integrated coastal zone management plan																																

Note 1: MAFF stand for the Ministry of Agriculture Forestry and Fisheries. Similarly, MoE for Ministry of Environment, MLIT for Ministry of Land, Infrastructure, Transport and Tourism, and MEXT for Ministry of Education, Culture, Sports, Science and Technology

Note 2: "◉" means the more direct and stronger relationships than "○", which shows relatively indirect or weak relationships

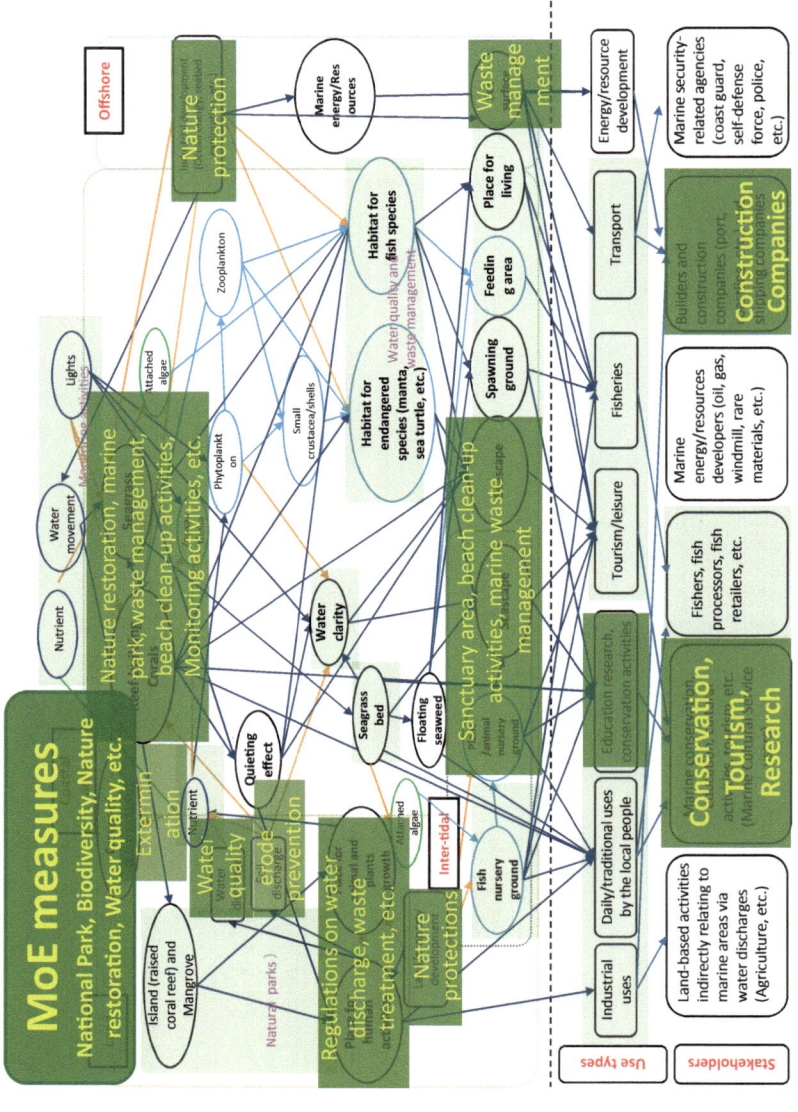

Fig. 2.3 Coverage of ecosystem structure, functions, and uses by policy measures under the charge of the Ministry of Environment (MoE)

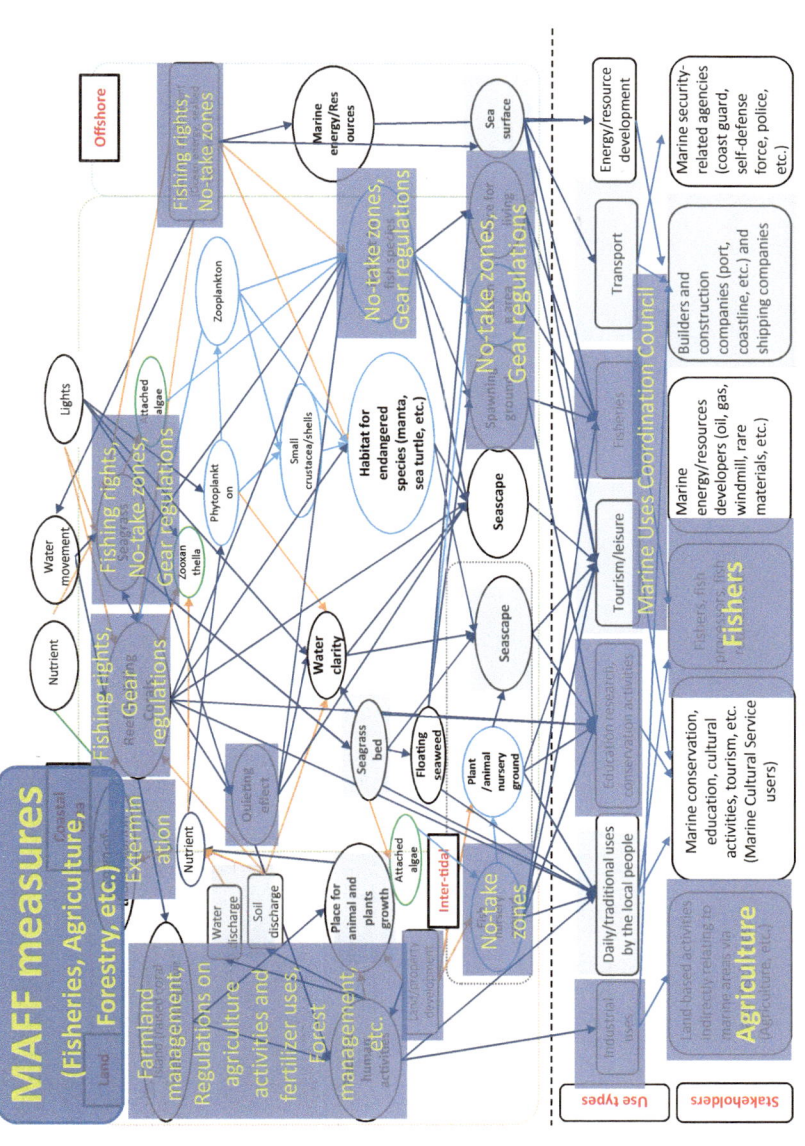

Fig. 2.4 Coverage of ecosystem structure, functions, and uses by policy measures under the charge of the Ministry of Agriculture, Forestry and Fisheries (MAFF)

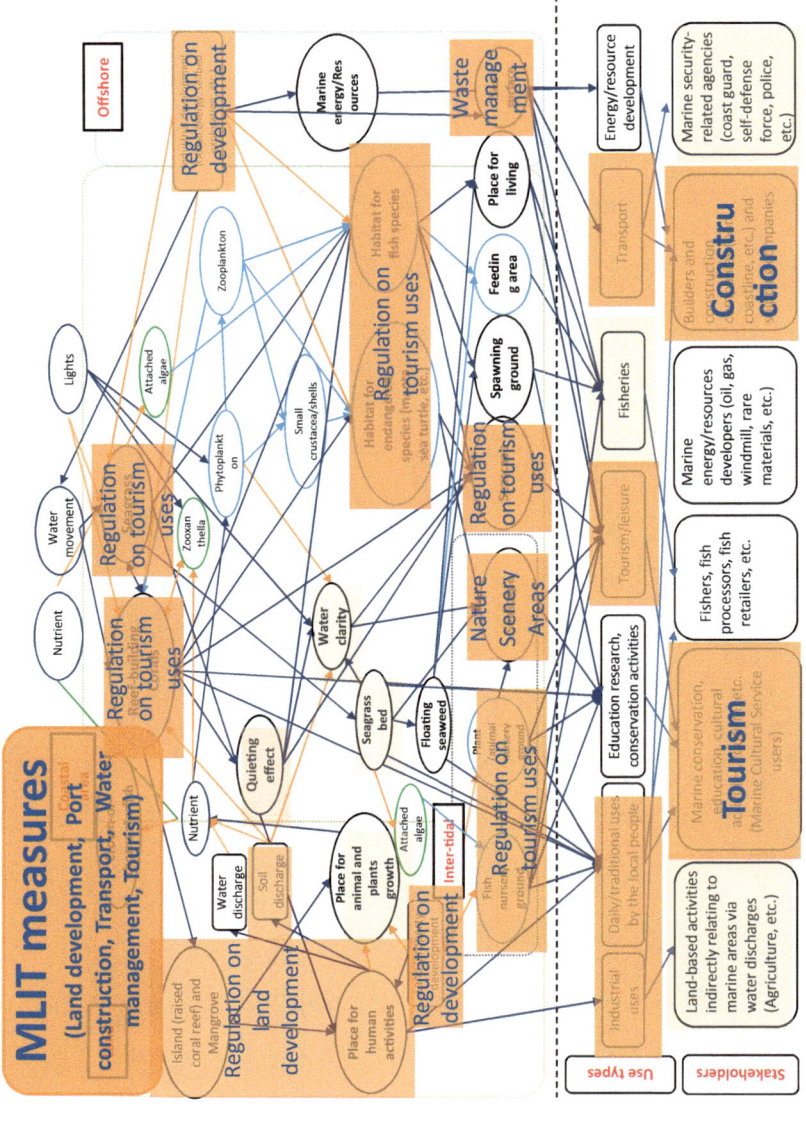

Fig. 2.5 Coverage of ecosystem structure, functions, and uses by policy measures under the charge of the Ministry of Land, Infrastructure and Transport (MLIT)

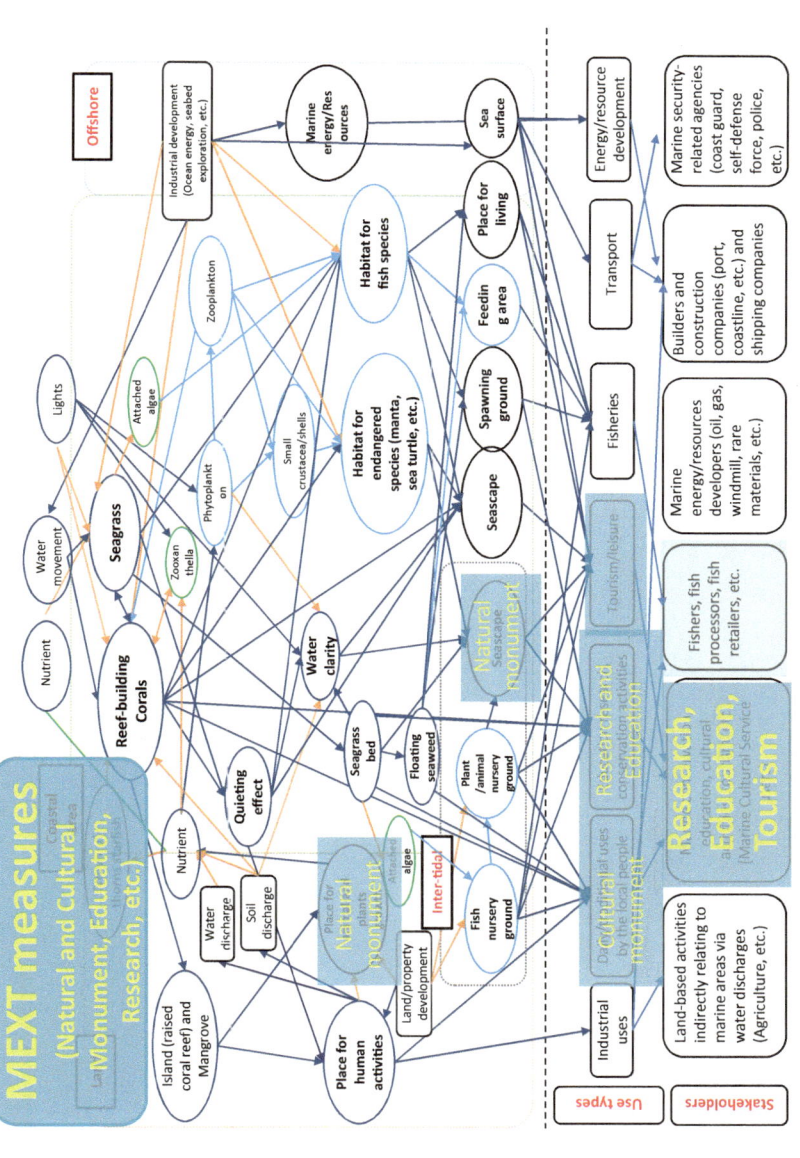

Fig. 2.6 Coverage of ecosystem structure, functions, and uses by policy measures under the charge of the Ministry of Education, Culture, Sports, Science and Technology (MEXT)

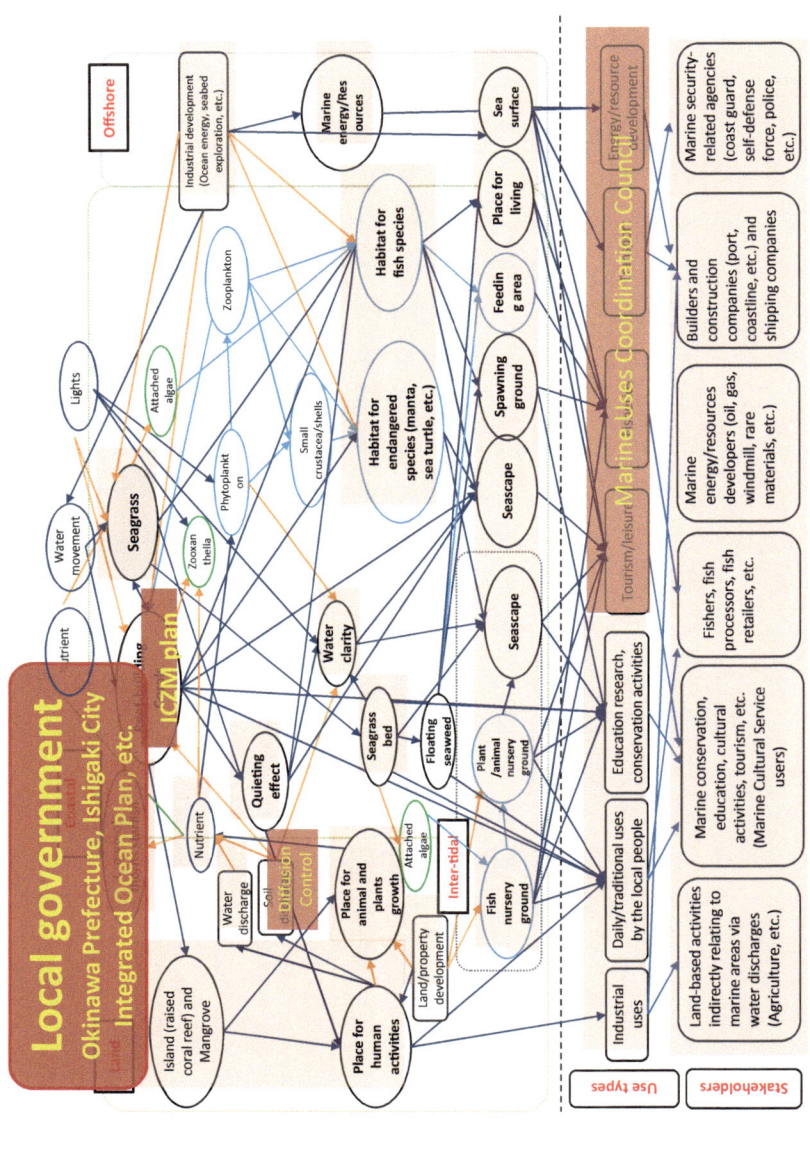

Fig. 2.7 Coverage of ecosystem structure, functions, and uses by policy measures under the charge of the Local Government (Okinawa Prefecture, Ishigaki City)

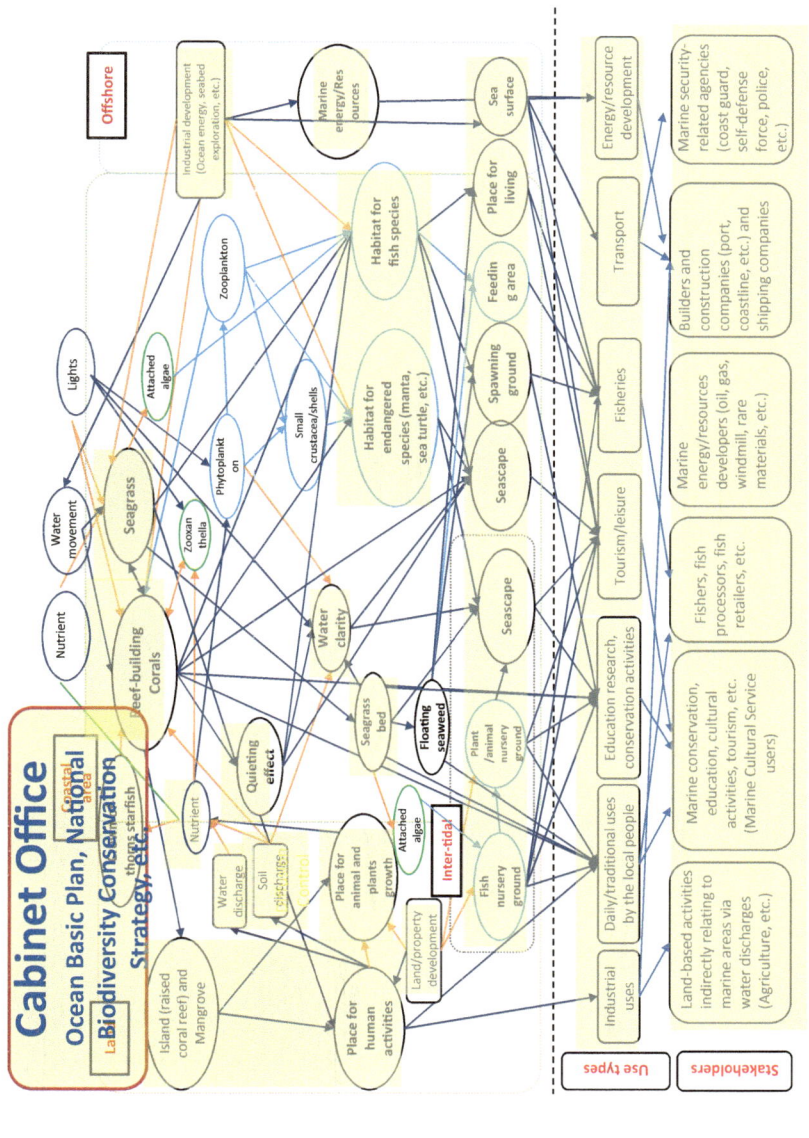

Fig. 2.8 Coverage of ecosystem structure, functions, and uses by policy measures under the charge of the Cabinet Office

2.4 Discussions

2.4.1 SES Schematic as a Boundary Object

This chapter tried to graphically summarize the main ecosystem structure, functions, use types and the stakeholders of the Sekisei Lagoon, and then link them to the various ministries and legal basis for the policy interventions. This is a genuine interdisciplinary work for understanding the coastal social-ecological systems (Armitage et al. 2017). This schematic can be used as a boundary object (Star and Griesemar 1989; Cash et al. 2003) to facilitate the knowledge exchanges among various stakeholders, to share the common understandings of the current situation, and to cocreate the innovative governance activities for the sustainable uses of the Sekisei Lagoon. As Reed et al. (2014) pointed out, inclusion of the stakeholders into the knowledge exchange scheme from the very early stage of a research project is important and effective option for the continued motivation and engagement by the stakeholders. Similarly, the knowledge exchange between researchers and decision-makers is important for effectively implementing the adaptive governance of the marine resources (Cvitanovic et al. 2015). Indeed, during the interviews about the policy interventions, the government officers and environmental policy experts often said to us that this SES schematic is useful to identify their administrative jurisdictions from a wider point of view, and to understand the interrelationships with other ministries or other sectors.

2.4.2 Integration of the Sectoral Policies and the Multilevel Governance

It is clearly shown in Fig. 2.2 that, the ecosystem structure and functions used by a certain stakeholder is closely connected to other structures and functions, which are then used by other sectors. Therefore, for example, environmental policy interventions for biodiversity conservation are also important for and effective to the sustainable fisheries (Friedman et al. 2018). This is one of the strongest messages made by the SES Schematic analysis.

Figures 2.2, 2.3, 2.3, 2.4, and 2.5 show that the majority of SES components are covered by some sectoral policy interventions by MoE, MAFF, MLIT, and MEXT. Therefore, all in all, sectoral policy interventions are covering the majority of the Sekisei Lagoon SES. This is the second strongest finding from this study. It means, the only remained task is the coordination from the viewpoint of the integrated ocean policy as a whole and the creation of the synergy effects. Theoretically, as Table 2.1 and Figs. 2.7 and 2.8 show, the local government and the Cabinet Office can cover most of the SES components. Therefore, to achieve the harmonization of uses and conservation (Article 2 of the Act), local government (Okinawa Prefecture, Ishigaki City), and the Cabinet Office (Headquarter for the Ocean Policy) can potentially play the central roles in the policy coordination and create the synergy effects. However, in reality, these organizations have smaller budgets, less staffs, and limited policy capacities compared to the sectoral ministries such as MoE, MAFF, MLIT, and MEXT. But they have advantages, as well. For example, local government has the local sense of the realities and the close connections to the local stakeholders. These are important and powerful advantages to codesign and co-implement the policy interventions. The Cabinet Office has the authority to coordinate other ministries, and it has the legal base to do that such as the Basic Ocean Plan of 2018 or the Marine Biodiversity Conservation Strategy, etc. Therefore, the more detailed analysis is required on how to design the multilevel governance framework for the effective ocean policy as a whole, and on the ideal role sharing and knowledge sharing scheme among the national government (including ministries), local government, local people, resource users, nongovernmental organizations, etc. (Jones 2014; Oyama 2017; Gerhardinger et al. 2018). Note that, as Makino et al. (2009) discussed, international organizations can sometimes play important roles to facilitate the multilevel and integrated governance within a country. Now, the Japanese government is planning to recommend this area to the UNESCO World Natural Heritage. The inscription to the Heritage List will bring additional effects for the ocean policy integration.

2.4.3 Next Step

This schematic is a qualitative expression of the interrelationships within the Sekisei Lagoon Social-Ecological Systems. Therefore, we cannot draw any lessons or implications about the cumulative effects or the trade-offs. Also, we cannot discuss the timescale issues or the magnitude of uncertainties or fluctuations within this schematic. In order to deal with these issues, quantitative model are needed. Also, we need the Geographical Information System (GIS) analysis to understand the spatial dynamics within the Sekisei Lagoon SES. However, taking the limitation of the research budget and human resources into account, it is neither realistic nor desirable to construct the detailed quantitative models for all the components of the Sekisei Lagoon SES. We need to set the priorities. In doing that, the SES Schematic can be utilized as a boundary object for local stakeholders, researchers, and decision-makers to discuss together to identify which part of the Sekisei Lagoon SES should be deeply analyzed by the detailed quantitative models. This is how the SES Schematic analysis and the quantitative modeling analysis can be linked in a transdisciplinary research project.

By incorporating the future climate scenarios to the SES Schematic, we can discuss the potential effects to both the ecological and social system. Figure 2.9 is the preliminary results of such exercise. As experienced in the Great Barrier Reef, the climate change will trigger the mass breaching and death of the coral reef, which then lead to the cross-scale effects to the larger ecosystem (Stuart-Smith et al. 2018). The Sekisei Lagoon is the same. In 2016, a large-scale mass breaching were observed

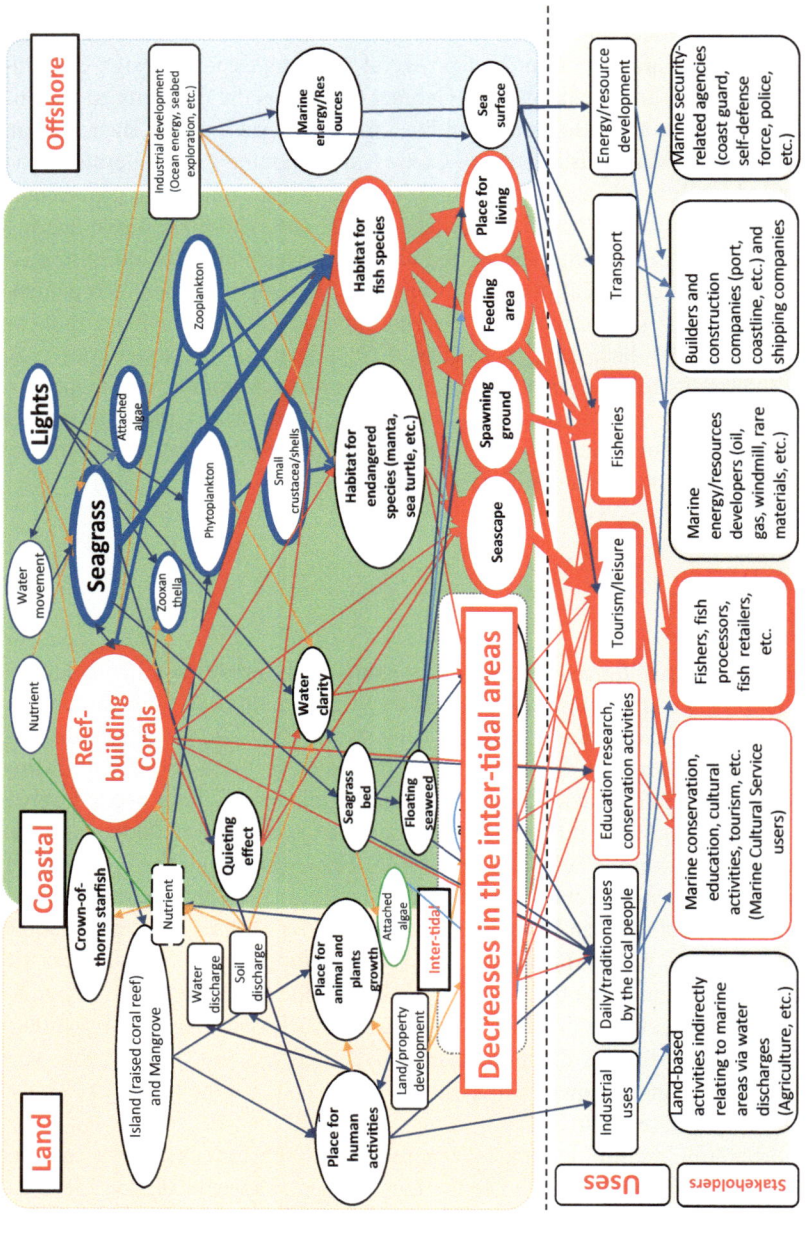

Fig. 2.9 Possible effects from climate change to the Sekisei Lagoon SES. Thick red arrows/circles mean the negative effects, and thick blue means the positive effects

in Okinawa Prefecture because of the extraordinary high temperature and the less typhoon visits, which led to the less water mixing. According to the official monitoring results by the Ministry of Environment, mass breaching was observed at 82–90% of the total area of the Sekisei Lagoon (Ministry of Environment 2018). If the climate change continues, the area of the coral reefs will be shrunken, and which leads to the decrease in the size of coral fish habitat Nanami et al. 2013), which then leads to the decrease in the spawning stock biomass of important fisheries resources (Nanami et al. 2015). Also, the higher water temperature leads to the changes in the spawning cycle, which presumably gives negative impact for spawning schooling (Ohta and Ebisawa 2017). These will give negative impacts to the fisheries and tourism sector. On the other hand, the higher water temperature positively effects to the seagrass bed at the coastal areas, which then positively effects to the fish biomass as the habitat for some species. Also, intertidal area will be shrunken by the sea level rise, which will again give negative effects to the traditional uses by the local people, as well as the environmental education, research and conservation activities. The cumulative effects to the total well-being and inclusive wealth (Ikeda and Managi 2018) are to be analyzed, but the SES Schematic is again a useful tool to share the knowledge with wide-ranging stakeholders, discuss about the vulnerable part of the SES, and codesign the possible future scenarios (Saito et al. 2018) or the adaptation plans.

Finally, after the development of the Sekisei Lagoon Social-Ecological Systems Schematic, the coauthor found that this type of co-research activity participated widely from both the natural and social sciences was a good opportunity for researchers to build a common sense at the larger conceptual level, and to create the knowledge base for working closely. Natural scientists can understand the links between their disciplinary study topics and the real society or the legal basis. Similarly, as discussed earlier, social scientists realized how the different stakeholders are interlinked at the ecosystem level. The coauthor also believe the developing process of the SES schematic will be utilized as the education program for students or the early career researchers who would like to conduct the interdisciplinary researches in the future.

Acknowledgments This study was funded by the Environment Research and Technology Development Fund (Strategic R&D Category) of Ministry of the Environment, Japan, "Predicting and Assessing Natural Capital and Ecosystem Services (PANCES)".

References

Armitage D, Charles A, Berkes F (eds) (2017) Governing the coastal commons: communities, resilience and transformation. Routledge, Oxford

Ban NC, Mills M, Tam J, Hicks CC, Klain S, Stoeckl N, Bottrill MC, Levine J, Pressey RL, Satterfield T, Chan KM (2013) A social-ecological approach to conservation planning: embedding social considerations. Front Ecol Environ 11(4):194–202

Berkes F, Armitage D, Ibarra MA, Charles T, Loucks L, Graham J, Seixas C, Makino M, Satria A, Abraham J (2014) Guidelines for analysis of social-ecological systems. http://www.communityconservation.net/ses-guidelines/. Accessed 18 Sept 2018

Bodin O (2017) Collaborative environmental governance: achieving collective action in social-ecological systems. Science 357:eaan114

Cash DW, Clark WC, Alcock F, Dickson NM, Eckley N, Guston DH, Jäger J, Mitchell RB (2003) Knowledge systems for sustainable development. Proc Natl Acad Sci 100(14):8086–8091

Cicin-Sain B, Knecht RW (1998) Integrated coastal and ocean management: concepts and practices. Island Press, Washington, DC

Cvitanovic C, Hobday AJ, van Kerkhoff L, Wilson SK, Dobbs K, Marshall NA (2015) Improving knowledge exchange among scientists and decision-makers to facilitate the adaptive governance of marine resources: a review of knowledge and research needs. Ocean Coast Manage 112:25–35

Diaz S et al (2018) Assessing nature's contributions to people: recognizing culture, and diverse sources of knowledge, can improve assessments. Science 359:270–272

Dunn WN (2016) Public policy analysis, 5th edn. Routledge, London

Ehler C (2014) A guide to evaluating marine spatial plans (IOC manuals and guides 70). UNESCO, Paris

Friedman K, Garcia SM, Rice J (2018) Mainstreaming biodiversity in fisheries. Mar Policy 95:209–220

Gerhardinger LC, Gorris P, Gonçalves LR, Herbst DF, Vila-Nova DA, De Carvalho FG, Glaser M, Zondervan R, Glavovic BC (2018) Healing Brazil's Blue Amazon: the role of knowledge networks in nurturing cross-scale transformations at the frontlines of ocean sustainability. Front Mar Sci 4:395

Guneroglu A (2015) Special issue: third international symposium on integrated coastal zone management (ICZM): towards sustainable coasts- "recent developments and advancements". Ocean Coast Manage 118:97–316

Hori J, Tajima H, Makino M (2017) The analysis of stakeholders' interests in coral reef ecosystems and their services-a case study on the Sekisei Lagoon. J Coast Zone Stud 30(2):61–73. (in Japanese)

Howard M, Davidson J, Lockwood M, Hockings M, Kriwoken L, Allchin R (2013) Climate change, scenarios and marine biodiversity conservation. Mar Policy 38:438–446

Ikeda S, Managi S (2018) Future inclusive wealth and human Well-being in regional Japan: projections of sustainability indices based on shared socioeconomic pathways. Sustain Sci. https://doi.org/10.1007/s11625-018-0589-7

Jones PJS (2014) Governing marine protected Aras: resilience through diversity. Routledge, New York

Kittinger JN, Koehn JZ, Le Cornu E, Ban NC, Gopnik M, Armsby M, Brooks C, Carr MH, Cinner JE, Cravens A, Erikson A, Finkbeiner EM, Foley MM, Fujita R, Gelcich S, St Martin K, Prahler E, Reineman DR, Shackeroff J, White C, Caldwell MR, Crowder LB (2014) A practical approach for putting people in ecosystem-based ocean planning. Front Ecol Environ 12(8):448–456

Lou X, Li Y, Chen F (2017) Coral reef restoration in Sekisei Lagoon, Okinawa, Japan. In: Guillotreau P, Bundy A, Perry I (eds) Global challenge in marine system: integrating natural, social and governing responses. Routledge, Oxon, pp 282–294

Makino M (2011) Fisheries management in Japan: its institutional features and case studies. Springer, Dordrecht

Makino M, Matsuda H, Sakurai Y (2009) Expanding fisheries co-management to ecosystem-based management: a case in the Shiretami world natural heritage, Japan. Mar Policy 33:207–214

Makino M, Hori M, Hori J, Nanami A, Tojo A, Tajima H (2018) Understanding the integrated policy for harmonizing the marine ecosystem conservation and sustainable uses: a case of Sekisei lagoon, Japan. Conference paper for OCEANS'18 MTS/IEEE Kobe/Techno-Ocean 2018, 28–31 May, Kobe

Ministry of Environment (2018) Press release: the monitoring results of the coral reef breaching phenomena around the Amami Island, Okinawa Island, Sekisei Lagoon and the Iriomote Island. http://www.env.go.jp/press/105831.html (in Japanese)

Nanami A, Sato T, Takebe T, Teruya K, Soyano K (2013) Microhabitat association in white-streaked grouper Epinephelus ongus: importance of Acropora spp. Mar Biol 160(6):1511–1517

Nanami A, Ohta I, Sato T (2015) Estimation of spawning migration distance of the white-streaked grouper (Epinephelus ongus) in an Okinawan coral reef system using conventional tag-and-release. Environ Biol Fish 98(5):1387–1397

Ohta I, Ebisawa A (2017) Inter-annual variation of the spawning aggregations of the white-streaked grouper Epinephelus ongus, in relation to the lunar cycle and water temperature fluctuation. Fish Oceanogr 26(3):350–363

Ostrom E (2009) A general framework for analyzing sustainability of social-ecological systems. Science 325:419–422

Oyama K (2017) Concept of governance in the social-ecological system theory: IPBES, Ostrom, and public governance. Keio Univ Legal Stud 90(3):1–31. (in Japanese)

Reed MS, Stringer LC, Fazey I, Evely AC, Kruijsen JHJ (2014) Five principles for the practice of knowledge exchange in environmental management. J Environ Manag 146:337–345

Saito O, Kamiyama C, Hashimoto S, Matsui T, Shoyama K, Kabaya K, Uetake T, Taki H, Ishikawa Y, Matsushita K, Yamane F, Hori J, Ariga T, Takeuchi K (2018) Co-design of national-scale future scenarios in Japan to predict and assess natural capital and ecosystem services. Sustain Sci. https://doi.org/10.1007/s11625-018-0587-9

Sakamoto S (2018) Ocean policy and ocean Laws in Japan. Shinzansha, Tokyo

Schultz L et al (2015) Adaptive governance, ecosystem management, and natural capital. PNAS 112:7369–7374

Star SL, Griesemar JR (1989) Institutional ecology, 'transi-tions' and boundary objects: amateurs and professionals in Berkeley's Museum of Vertebrate Zoology, 1907-1939. Soc Stud Sci 19(3):387–420

Stuart-Smith RD, Brown CJ, Ceccarelli DM, Edgar G (2018) Ecosystem restructuring along the Great Barrier Reef following mass coral bleaching. Nature 560:92–96

Sugimoto A (2016) Fish as a 'bridge' connecting migrant fishers with the local community: findings from Okinawa, Japan. Maritime. Studies 15:5

Tiller RG, Kok J-LD, Vermeiren K, Thorvaldsen T (2017) Accountability as a Governance Paradox in the Norwegian Salmon Aquaculture Industry. Front Mar Sci 4:71

Weimer DL, Vining AR (2017) Policy analysis: concepts and practice, 6th edn. Routledge, New York

Chapter 3
How to Engage Tourists in Invasive Carp Removal: Application of a Discrete Choice Model

Kota Mameno, Takahiro Kubo, Yasushi Shoji, and Takahiro Tsuge

Abstract Invasive alien species management requires public participation to overcome a lack of human and financial resources in management; however, little is known about the demand for public participation in invasive alien species management. To address this knowledge gap, the present study evaluated demand for management of invasive carp, which is one of the worst but publicity invasive species worldwide. A choice experiment survey was conducted in Amami Oshima Island, Japan to quantify tourists' demand for participating in invasive carp removal in nature-based tourism, and to evaluate the impact of ecological information provision on their preference. The results show most tourists would avoid participating in carp removal activities as a tour option without any financial discounts; however, over 35.2% of tourists were willing to work for carp removal, based on their own motivations. We also found that ecological information encouraged tourists to participate in tours that included carp removal activities. Incorporation of invasive alien species management in nature-based tourism can enhance the economic benefits for local tourism industries. Our findings indicate that tourists could play an important role in invasive alien species management by compensating for a lack of human and financial resources in management.

K. Mameno (✉)
Graduate School of Agriculture, Hokkaido University,
Sapporo, Japan

Center for Environmental Biology and Ecosystem Studies, National Institute for
Environmental Studies, Tsukuba, Japan
e-mail: tls159-red@eis.hokudai.ac.jp

T. Kubo
Center for Environmental Biology and Ecosystem Studies, National Institute for
Environmental Studies, Tsukuba, Japan

Y. Shoji
Research Faculty of Agriculture, Hokkaido University, Sapporo, Japan

T. Tsuge
Faculty of Economics, Konan University, Kobe, Japan

O. Saito et al. (eds.), *Managing Socio-ecological Production Landscapes and
Seascapes for Sustainable Communities in Asia*, Science for Sustainable
Societies, https://doi.org/10.1007/978-981-15-1133-2_3

Keywords Asian carp · Discrete choice model · Information provision · Public engagement

3.1 Introduction

A lack of human and financial resources represents a bottleneck to invasive alien species (IAS) management all over the world, although it is widely recognized that IAS causes biodiversity loss (Didham et al. 2005; Sala et al. 2000). IAS have substantially damaged natural resource-based industries, such as agriculture, forestry, and fishery (Pejchar and Mooney 2009; Pimentel et al. 2005). For some governments and environmental managers, access to human and financial resources in long-term represents one of the biggest challenges for effective IAS management (Gardener et al. 2010; Simberloff et al. 2005).

IAS management programs have considered that public involvement is an essential tool to succeed, since it could overcome the management resources shortage, such as a lack of human and financial resources (Dunn et al. 2018; Gaertner et al. 2016; McNeely 2001). For example, common sun skinks (*Eutropis multifasciata*) were removed by tourists on Green Island, leading to a decrease in skink numbers (Chao and Lin 2017). Based on these results, the Convention on Biological Diversity (CBD) mentioned that citizens were one of the most important players in IAS management (CBD 2014).

In spite of a general agreement on the importance of public involvement toward the success of IAS management, such involvement is still difficult to achieve. Previous literature noted that public apathy toward IAS management represents one of the biggest barriers to public involvement. For example, people who underestimate IAS impacts in Japan tended not to support IAS management (Akiba et al. 2012; Mameno et al. 2017). Thus, public education and information provision could be effective in engaging more people (Bremner and Park 2007; Marzano et al. 2015). Based on these suggestions, some governments have provided relevant information; however, public awareness and public involvement remains insufficient (Dunn et al. 2018). Recently, the focus shifted to indirect approaches, which comprise neither education nor information provision. For example, Morgan and Ho (2018) showed that good tasting carp meat encouraged invasive carp removal.

So far, many previous studies have addressed the public attitude concerning an IAS and the management thereof (Bremner and Park 2007; Mameno et al. 2017; Wald et al. 2018), as well as the estimation of social values from IAS management programs (Nunes and Van Den Bergh 2004; Roberts et al. 2018). However, little research has focused on how to encourage people to participate in IAS management activities. Thus, our study—which uses a tour that comprises carp removal activities—can provide new insights into public engagements in IAS management. This work also extends the knowledge of voluntary conservation approaches. A few studies have focused on such approaches (Durán-Medraño et al. 2017); however, to our knowledge, no studies have addressed tourists' attitudes concerning participation in conservation management.

The present study focuses on invasive common carp (*Cyprinus carpio*) management from the human dimension perspective. The invasive carp is one of the most invasive species, and is nominated as one of "100 of the world's worst invasive alien species" (Kopf et al. 2017; Lowe et al. 2000), since the carp damages the native ecosystem by consuming organisms that are a food resource for native species (Gilligan and Rayner 2007; Morgan and Hicks 2013). Thus, recent conservation literature paid attention to its management (Marshall et al. 2018; Thresher et al. 2014; Uchii et al. 2014); however, social science research on this issue is limited— for example, Morgan and Ho (2018). Interestingly, the carp is a domestic alien species in Japan—it is a native species on the main island, but non-native on some of the small islands (e.g., Amami Oshima Island, which is our research site). Since few people would recognize carp as an IAS in Japan, the sharing of ecological information could play an important role, as suggested by previous literature. Historically, IAS management, including invasive carp, was greatly dependent on chemical and biological methods (Zastrow 2018). Thus, our research contributes to the existing body of knowledge on effective invasive carp management through application of human dimension approaches (Jacobson and Duff 1998; McNeely 2001).

3.2 Study Background and Methods

3.2.1 Research Site

Our research site, Amami Oshima Island, Japan, is part of the Nansei Islands in the southern Japanese archipelago (28°19′N, 128°22′E). The island has the Japan's second largest mangrove forest, which plays an important role in biodiversity conservation (Lugo and Snedaker 1974; Polidoro et al. 2010). In particular, there are endangered and endemic species, such as Ryukyu Ayu (*Plecoglossus altivelis ryukyuensis*), around the mangrove forest (Kishino and Yonezawa 2013; Nishida 1988). Because of its rich biodiversity, the part of the island that includes the mangrove forest is a national park that is expected to become a Natural World Heritage Site. However, the island has substantial IAS concerns, and has been requested by UNESCO, through its designation process, to enhance its IAS management. Local government has attempted to remove common carp so as to improve its management in rivers (Fig. 3.1); however, the success has been limited because of resource constraints, among others.

Nature-based tourism is an important industry on the island. To date, most tourists have enjoyed nature-based and eco-friendly tours, such as canoeing in the mangrove forest and viewing wildlife, in spite of the potential invasive carp impacts on the ecosystem. The canoe tour attracts over 30,000 tourists annually (Kagoshima Prefecture 2019). It is the second most popular recreational activity in the island

Fig. 3.1 Invasive carp captured on Amami Oshima Island

(Ministry of the Environment 2017). Thus, balancing tourism development and bio-diversity conservation represents a significant challenge for local government. Based on this background, the present study evaluates tourists' demand for carp removal options as a canoe tour attribute, and discusses the possibilities to compensate for management resource constraints in invasive carp management through nature-based tourism.

3.2.2   Questionnaire Design

In this study, we used a choice experiment (hereinafter, "CE") on canoe tours. The CE approach is one of methods for analyzing preferences through hypothetical choices in a questionnaire survey; it has been applied by many previous studies to nature-based tourism (Kubo et al. 2019; Kubo and Shoji 2016). We used this approach to evaluate tourists' willingness to pay (WTP) for tour options as a means of promoting canoe tours. Using the CE can elicit not only the level of demand for the option of carp removal, but also assess other tour options, such as tour time and fee.

	Attribute	Levels
Table 3.1 Attributes and levels for profile design using the choice experiment	Carp removal option	Yes; No
	Tour time (min)	60; 90; 120; 150; 180
	Pick-up option	Yes; No
	Tour fee (JPY)	1000; 2500; 5000; 7500; 10,000

A distributed questionnaire of six pages was used for the CE valuation exercise. Respondents were asked to choose their preferred option from alternative tour scenarios with different combinations of tour options.

In CE studies, it is important to select attributes and levels to create hypothetical scenarios. Based on current tours and discussions with managers on the Amami Oshima Island, this study selected the following attributes and levels to design profiles and choice sets: *Carp Removal option, Tour Time, Pick-up option*, and *Tour Fee. Carp Removal option* refers to catching carp using fishing nets before and/or after the canoe tour. *Pick-up option* refers to picking up tourists at their accommodations and taking them to recreation sites. In the island, recreational sites are located far from towns and the airport. Thus, tourists have to drive themselves to recreational areas or use a bus service, which only operates every few hours. These four characteristics, and possible choices for each, are listed in Table 3.1.

The respondents were asked to choose from three tour options under the hypothetical scenario. The scenario implied that new attributes of canoe tours (i.e., *Carp Removal option* and *Pick-up option*) could be implemented, and the levels of existing attributes (i.e., *Tour Time* and *Tour Fee*) could be changed to encourage the use of canoe tours.

Considering the scenario, we designed profiles and choice sets based on D-efficiency. The D-efficient design is able to minimize the distribution of estimated parameter, which contributes to efficient parameter estimations (Huber and Zwerina 1996). To mimic actual tour choice situations (Haaijer et al. 2001), we created choice sets that consist of "not attending tour" as well as two selected profiles (an example of choice set is shown as Fig. 3.2); we then created six patterns of a questionnaire with five choice sets each, and provided each respondent with one randomly selected questionnaire.

In addition, two types of questionnaires were used to assess the impact of information provision, namely, to understand differences between the preferences of respondents who recognized carp as an IAS, and those who did not. The information that "*Carp are an IAS and cause serious damage to the native unique ecosystem on Amami Oshima Island; this option contributes to the conservation of ecosystems*" was provided to some respondents only, and not to the rest. Respondents without this ecological information could have knowledge of invasive carp impacts. However, our study investigated the impact of information provision by focusing on the difference in attitude toward preference for tours between respondents with information and without information.

	Tour 1	Tour 2	Non-attend the tours
Carp Removal option	Yes	No	
Tour Time (minutes)	60	150	
Pick-up option	Yes	No	
Tour Fee (JPY)	5000	2500	
Please circle the answer ⇒	↓ 1	↓ 2	↓ 3

Fig. 3.2 Example of a choice set using a choice experiment. In the choice experiment, we showed three profiles (two profiles with different levels for the four attributes, while another profile is nonattendance) to each respondent, and repeated this task five times

3.2.3 Data Collection

The tourist questionnaire survey was conducted with randomized distribution at the Amami Airport on Amami Oshima Island in August 2017. Nine hundred and twenty-four questionnaires were distributed to tourists, of which 343 questionnaires were returned by mail (the response rate was 37.1%); of these, 12 contained no answers to any of the choice experiment questions and were thus omitted from analysis. We ultimately used data of 331 respondents, and 1608 choice sets contained answers to all relevant questions. Of the about half of respondents ($n = 175$) were provided ecological information, and 68.0% of respondents were female ($n = 217$). The most represented age group was between 40 and 49 years of age ($n = 92$), followed by respondents between 30 and 39 years of age ($n = 84$).

3.2.4 Econometric Model

To analyze the tourists' preferences, we used a model based on random utility theory (McFadden 1974), which comprises an observable deterministic component and an unobservable random component. According to this model, each individual's utility (U_i) of alternative i can be described as a function of an observable component (V_i) and an unobservable random (error) component (ε_i):

$$U_i = f(V_i, \varepsilon_i) = V_i + \varepsilon_i \tag{3.1}$$

The alternative i was chosen by the individual if $U_i > U_j$ for all $j \neq i$. Thus, the probability that the respondent chooses the alternative i from choice set C is:

$$P_i = \Pr\left[U_i > U_j\right] \quad \forall j \neq i, \quad \forall j \in C$$
$$= \Pr\left[V_i + \varepsilon_i > V_j + \varepsilon_j\right] = \Pr\left[V_i - V_j > \varepsilon_j - \varepsilon_i\right] \tag{3.2}$$

The probability of choosing alternative i is able to be written the following equation if each ε_i is assumed to distribute with a type I extreme value distribution,

$$P_i = \frac{\exp(V_i)}{\sum_{j \in C} \exp(V_j)} \tag{3.3}$$

While we estimated the results using both a conditional logit (CL) and a random parameter logit (RPL) models, in the present study, the results of only the RPL models were represented. That was why the results of the RPL models were superior to those of the CL as followers. First, RPL models do not require fulfilling the independence of irrelevant alternatives (IIA) condition, which is required by the conditional logit model. In addition, previous studies showed that tourists had preference heterogeneity (Kubo et al. 2019; Mejía and Brandt 2015) for nature-based tours, which is accommodated by RPL models.

According to Hensher et al. (2015), in RPL models, V_{nit} on the Eq. 3.3 is showed the following form using the individual n, choosing alternative i and period t:

$$V_{nit} = \beta_n' x_{nit}$$
$$\beta_n = \beta + \Delta Z_n + \Gamma v_n \tag{3.4}$$

where β indicates the population mean of the coefficient of random parameters; ΔZ_n indicate observed preference heterogeneity; on the other hand, unobserved preference heterogeneity is showed in Γv_n; x_{nit} is the attributes in choice set which respondents are asked. This model can incorporate unobserved and observed preference heterogeneity through the random terms in the distributions of parameters (Hensher et al. 2015). We are able to calculate the expected probability using multivariate probability density function of β:

$$E_n = \int P_n(\beta) \cdot f(\beta | \Omega) d\beta = \int \left[\frac{\exp(\beta_n' x_{nit})}{\sum_{j \in C} \exp(\beta_n' x_{nit})} \cdot f(\beta)\right] d\beta \tag{3.5}$$

The parameter is estimated by simulated maximum likelihood techniques that maximize the log-likelihood function, because the integral of estimating this model does not have a closed form (McFadden and Train 2000; Train 2009).

In present study, we used *Carp Removal option*, *Pick-up option*, *Tour Time* and *Tour Time* squared (*Tour Time²*), *Tour Fee*, and an alternative-specific constant (ASC) as explanatory variables: x_{nit}, and as observed preference heterogeneity, we incorporate the dummy variable of information provision that is if respondents were provided ecological information, the variable is set to 1. The dummy variable of information provision was incorporated as a cross-term with other attributes except for *Tour Fee* (i.e., *Carp Removal option, Pick-up option, Tour Time* and *Tour Time²*). By following existing literature in applied economics, preference for the *Tour Fee* (i.e., money) was supposed not to be affected by the information (e.g., Das et al. 2009; Hole and Kolstad 2012; Layton and Brown 2000; Morey and Rossmann 2003; Revelt and Train 1998; Scarpa et al. 2008). The ASC variable was set to 1 if respondents chose not to attend the canoe tour (i.e., selected choice 3 in choice set), and the ASC variable was set to 0 if respondents selected to attend the canoe tour (i.e., selected choice 1 or 2 in choice set). This model needed to make assumptions about the distributional form of the coefficients of each attribute. Except for *Tour Fee*, all variables were assumed to be normally distributed.

In addition, the marginal willingness to pay (MWTP) is able to be calculated by using the estimates for each coefficient as following equation (Haab and McConnell 2003):

$$\text{MWTP} = \frac{\beta_{\text{attribute}}}{\beta_{\text{Tour Fee}}} \tag{3.6}$$

Note that the WTP for both *Carp Removal option* and *Pick-up option* attributes was calculated by doubling the MWTP since the attributes were effect-coded (Louviere et al. 2000).

3.3 Results

Table 3.2 presents the estimated results of the RPL model. We applied the effect code to the dummy variables concerning *Carp Removal option* and *Pick-up option*. Thus, the relative values of coefficient are important instead of the absolute values. Table 3.2 shows the mean of coefficients, standard deviations of the random coefficients, and the effect of ecological information on each random parameter variables in the RPL models, with their standard errors. The standard deviations for each of the random coefficients indicate the heterogeneity of individual preferences relative to the preference of the population.

Table 3.2 indicates that the mean of the *Carp Removal option* coefficient is significantly negative (Coefficient = −0.366; SE = 0.131) at the 99.9% level: *Carp Removal option* decreased the probability of choosing canoe tours. However, standard deviations for *Carp Removal option* are also statistically significant: there is preference heterogeneity for *Carp Removal option* (Fig. 3.3). The mean of the respondents' WTP for *Carp Removal option* was −988 JPY (100 JPY = about 0.9 USD and 0.8 EURO in November 2019). The other variables' coefficients are also significant at over 95% levels, except for ASC. The coefficient of *Pick-up option*

Table 3.2 Estimation results using the random parameter logit model

	Random parameter model	
Variable	Mean of coefficient (S.E)	Standard deviation of coefficient (S.E)
Random and nonrandom parameters in utility function		
Carp removal option	−0.366 (0.131)***	1.01 (0.112)***
Tour time $*10^{-1}$	0.395 (0.136)***	0.00112 (0.0374)
Tour time^2 $*10^{-4}$	−1.52 (0.586)***	0.986 (0.117)***
Pick-up option	0.405 (0.110)***	0.719 (0.107)***
Tour fee $*10^{-3}$	−0.741 (0.0514)***	–
Non-attendance (alternative-specific constant)	−1.25 (0.818)	2.52 (0.329)***
The effect of ecological information on each random parameter variable		
Carp removal option	0.403 (0.176)**	
Tour time $*10^{-1}$	−0.0283 (0.177)	
Tour time^2 $*10^{-4}$	−0.324 (0.768)	
Pick-up option	0.0138 (0.146)	
Non-attendance (alternative-specific constant)	0.0273 (1.08)	
Number of observations	1608	
Log likelihood	−1243.10	

Simulated maximum likelihood was conducted using Halton draws with 1000 replications
Please, see Train 2003 for details on Halton draws
Carp Removal option: availability of options for carp capture; *Pick-up option*: collection and transport to the recreational site by guides; *Tour Time*: duration of tour (h); *Tour Time*^{2}: *Tour Time* squared; *Tour Fee* (JPY). *Carp Removal option* and *Pick-up option* were applied effect coding; *Tour Fee*, *Tour Time*, and *Tour Time*^{2} were normalized
$***p < 0.001$, $**p < 0.01$, $*p < 0.05$

and *Tour Time* is positive, on the other hand, the coefficients of *Tour Time*^{2}, *Tour Fee* are negative. While standard deviation of *Tour Time* coefficient is not significant, the other variables' standard deviation of coefficients is also significant: there is preference heterogeneity.

Table 3.2 also shows the effect of ecological information provision, and they are not significant except for *Carp Removal option*. The coefficient of the effect of ecological information on *Carp Removal option* is significantly positive (Coefficient = 0.403; SE = 0.176), namely, information provision dramatically increased the probability of choosing canoe tours including carp removal (Fig. 3.3). The coefficient of *Carp Removal option* for respondents who were provided with information could be calculated through addition of the coefficient of *Carp Removal option* and the effect of ecological information on *Carp Removal option*, which had a positive result (Coefficient = 0.0368). Therefore, the utility of respondents had ecological information became positive if the option for carp removal was added to canoe tours, and the mean WTP for *Carp Removal option* by respondents who had ecological information was estimated at 99.3 JPY.

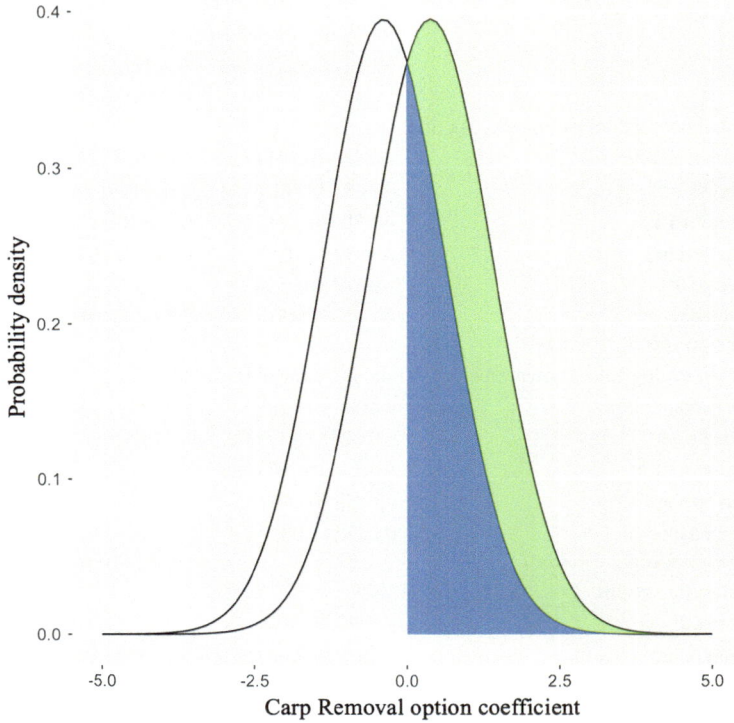

Fig. 3.3 Results of the random parameter logit model. The range of "Blue" indicates the probability density of respondents without ecological information that answered positive to *Carp Removal option*. On the other hand, the range of "Green" indicates that providing ecological information made the increasing probability density of respondents who answered positive to *Carp Removal option*. Even though respondents had no ecological information, a part of the respondents (Blue) answered positive to *Carp Removal option*. When respondents were provided ecological information, over half of the respondents (Blue and Green) answered positive to *Carp Removal option*

3.4 Discussion and Conclusion

Public participation is an essential enhancement to IAS management, since it can compensate for the shortages of human and financial resources (Jenkins 1999; Lepczyk 2005; Trumbull et al. 2000); however, few studies have addressed the potential roles of tourists, although most of them visit wilderness areas such as national parks. We applied a discrete choice model to invasive carp management on Amami Oshima Island, Japan, to quantify tourists' demand for participating in the carp removal program as a tour option. The results indicate that 64.2% of tourists who were not provided ecological information, about 100 tourists would avoid participating in carp removal activities as a tour option without any financial discounts when they have no ecological information (Table 3.2). In other words, on average, tourists with no information are indifferent to whether they participate in carp removal activities if they received a 988 JPY discount for a tour. This finding suggests that an additional IAS management option could decrease the number of tourists without any discounts, although preference heterogeneity should be discussed in more detail, which is described as follows. On the other hand, given the preference heterogeneity concerning the carp removal option by RPL models, 35.8% of tourists were willing to work for carp removal owing to their own motivations. Some tourists regarded carp removal as a tour activity, even when they were not provided with any ecological information (Fig. 3.3, Blue). This implies that a part of their preference heterogeneity could be derived from their original knowledge of and interests in nature and invasive management. This result also supports the view that tourist involvement could help to overcome the cost challenges of IAS management.

Although few studies have addressed IAS management as a tour activity, our finding is contrary to previous studies which have shown positive public support for IAS management. For example, Nishizawa et al. (2006) estimated the WTP for an eradication program of black bass (*Micropterus salmoides* and *Micropterus dolomieu*) as loading tax in Japan at about 2000 JPY. McIntosh et al. (2010) also showed that the WTP for invasive fish as a donation was about 50 USD. Mejía and Brandt (2015) showed that tourists had a positive WTP for a tour that enhanced IAS management. This contradiction indicates that the public, including tourists, has a motivation for financial support of IAS management; however, most of them are not willing to participate in IAS management, which suggests that human resources could become insufficient compared to financial resources in general.

How do we then engage the public in IAS management? The present study supports the views of previous literature by showing that the provision of ecological information affects tourists' preference for carp removal activities in a tour (Bremner and Park 2007; Marzano et al. 2015). Tourists who received ecological information have a positive WTP (99.3 JPY) for participation in carp removal activities, which is 1087 JPY more than tourist who had no ecological information. In other words, provision of ecological information enables tour operators to receive a tour premium for including IAS management opportunities. This suggests that information provision leads to a win-win situation by enhancing IAS management while satisfying tourists' demands and increasing tour operators' profits. Recent tourism literature highlights that balancing conservation and the local economy through nature-based tourism is becoming increasingly important (Kubo et al. 2019). We demonstrated that this approach can even succeed in IAS management by sharing ecological knowledge with the public, as described in conservation literature (Akiba et al. 2012; Bremner and Park 2007; Dunn et al. 2018; Mameno et al. 2017; Marzano et al. 2015).

The other attributes concerning canoe tour design are also important for tourists' decision-making and their involvement in carp removal activities. As shown in Table 3.2, tourists prefer to participate in tours of intermediate duration (i.e., about 120 min), and in tours that offer a pick-up option. These findings indicate that tourists are more encouraged to participate in a tour by adjusting the levels of the attributes of tour time and pick-up options, regardless of a carp removal option. Previous work by Morgan and Ho (2018) indicated that indirect approaches, which are not directly related with IAS management, are important. The present findings support their view of such indirect approaches, even in invasive carp removal management, and outline a new approach based on the use of nature-based tourism.

Our research site, Amami Oshima Island, is expected to be designated as a Natural World Heritage Site. Thus, IAS management has been paid more attention recently; however, a lack of resources has limited the implementation of IAS management. Our findings indicate that nature-based tourism which combined with the sharing of ecological knowledge with stakeholders, as well as indirect approaches, achieves sustainable IAS management.

Acknowledgement We acknowledge financial support from the Japan Society for the Promotion of Science (No. 16K00697), and the Ministry of the Environment, Japan (the Environmental Economics and Policy Study, and ERTDF (S-15: Predicting and Assessing Natural Capital and Ecosystem Services (PANCES)). We would like to express our gratitude to Oki, K, and the tourists who responded to the survey. We also thank an editor and reviewers for helpful comments on the previous manuscript.

References

Akiba H, Miller CA, Matsuda H (2012) Factor influencing public preference for raccoon eradication plan in Kanagawa, Japan. Hum Dimens Wildl 17:207–219

Bremner A, Park K (2007) Public attitudes to the management of invasive non-native species in Scotland. Biol Conserv 139:306–314

CBD (2014) Pathways of introduction of invasive species, their prioritization and management. Secretariat of the Convention on Biological Diversity, Montréal

Chao RF, Lin TE (2017) Effect of citizen action on suppression of invasive alien lizard population: a case of the removal of Eutropis multifasciata on Green Island, Taiwan. Appl Ecol Environ Res 15:1–13

Das C, Anderson CM, Swallow SK (2009) Estimating distributions of willingness to pay for heterogeneous populations. South Econ J 75:593–610

Didham RK, Tylianakis JM, Hutchison MA, Ewers RM, Gemmell NJ (2005) Are invasive species the drivers of ecological change? Trends Ecol Evol 20:470–474

Dunn M, Marzano M, Forster J, Gill RMA (2018) Public attitudes towards "pest" management: perceptions on squirrel management strategies in the UK. Biol Conserv 222:52–63

Durán-Medraño R, Varela E, Garza-Gil D, Prada A, Vázquez MX, Soliño M (2017) Valuation of terrestrial and marine biodiversity losses caused by forest wildfires. J Behav Exp Econ 71:88–95

Gaertner M, Larson BMH, Irlich UM, Holmes PM, Stafford L, van Wilgen BW, Richardson DM (2016) Managing invasive species in cities: a framework from Cape Town, South Africa. Landsc Urban Plan 151:1–9

Gardener MR, Atkinson R, Rentería JL (2010) Eradications and people: lessons from the plant eradication program in Galapagos. Restor Ecol 18:20–29

Gilligan DM, Rayner T (2007) The distribution, spread, ecological impacts and potential control of carp in the upper Murray River. NSW Department of Primary Industries, Cronulla

Haab T, McConnell K (2003) The econometrics of non-market valuation. Edward Elgar, Northampton

Haaijer R, Kamakura W, Wedel M (2001) The 'no-choice' alternative in conjoint choice experiments. Int J Mark Res 43:93–106

Hensher D, Rose JM, Greene W (2015) Applied choice analysis. Cambridge University Press, Cambridge

Hole AR, Kolstad JR (2012) Mixed logit estimation of willingness to pay distributions: a comparison of models in preference and WTP space using data from a health-related choice experiment. Empir Econ 42:445–469

Huber J, Zwerina K (1996) The importance of utility balance in efficient choice designs. J Mark Res 33:307–317

Jacobson SK, Duff MD (1998) Training idiot savants: the lack of human dimensions in conservation biology. Conserv Biol 12:263–267

Jenkins EW (1999) School science, citizenship and the public understanding of science. Int J Sci Educ 21:703–710

Kagoshima Prefecture (2019) Trends in Amami Islands Tourism 2018. http://www.pref.kagoshima.jp/aq01/chiiki/oshima/chiiki/zeniki/oshirase/documents/38010_20190422104935-1.pdf (In Japanese)

Kishino T, Yonezawa T (2013) Seasonal distribution of Ryukyu-ayu Plecoglossus altivelis ryukyuensis in the Katoku River, Amami-Oshima Island, southern Japan. Jpn J Ichthyol 60:91–101

Kopf RK, Nimmo DG, Humphries P, Baumgartner LJ, Bod M, Bond NR, Byrom AE, Cucherousset J, Keller RP, King AJ (2017) Confronting the risks of large-scale invasive species control. Nat Ecol Evol 1:172

Kubo T, Shoji Y (2016) Demand for bear viewing hikes: implications for balancing visitor satisfaction with safety in protected areas. J Outdoor Recreat Tour 16:44–49

Kubo T, Mieno T, Kuriyama K (2019) Wildlife viewing: the impact of money-back guarantees. Tour Manag 70:49–55

Layton DF, Brown G (2000) Heterogeneous preferences regarding global climate change. Rev Econ Stat 82:616–624

Lepczyk CA (2005) Integrating published data and citizen science to describe bird diversity across a landscape. J Appl Ecol 42:672–677

Louviere JJ, Hensher DA, Swait JD (2000) Stated choice methods: analysis and applications. Cambridge University Press, Cambridge

Lowe S, Browne M, Boudjelas S, De Poorter M (2000) 100 of the world's worst invasive alien species: a selection from the global invasive species database. Invasive Species Specialist Group, Auckland

Lugo AE, Snedaker SC (1974) The ecology of mangroves. Annu Rev Ecol Syst 5:39–64

Mameno K, Kubo T, Suzuki M (2017) Social challenges of spatial planning for outdoor cat management in Amami Oshima Island, Japan. Glob Ecol Conserv 10:184–193

Marshall J, Davison AJ, Kopf RK, Boutier M, Stevenson P, Vanderplasschen A (2018) Biocontrol of invasive carp: risks abound. Science 359:877–877

Marzano M, Dandy N, Bayliss HR, Porth E, Potter C (2015) Part of the solution? Stakeholder awareness, information and engagement in tree health issues. Biol Invasions 17:1961–1977

McFadden D (1974) The measurement of urban travel demand. J Public Econ 3:303–328

McFadden D, Train K (2000) Mixed MNL models for discrete response. J Appl Econ 15:447–470

McIntosh CR, Shogren JF, Finnoff DC (2010) Invasive species and delaying the inevitable: valuation evidence from a national survey. Ecol Econ 69:632–640

McNeely J (2001) The great reshuffling. Human dimensions of invasive alien species. IUCN, Gland/Cambridge

Mejía CV, Brandt S (2015) Managing tourism in the Galapagos Islands through price incentives: a choice experiment approach. Ecol Econ 117:1–11

Ministry of the Environment (2017) Amami Islands national park designation and construction plan. https://www.env.go.jp/park/amami/intro/files/plan_02.pdf (In Japanese)

Morey E, Rossmann KG (2003) Using stated-preference questions to investigate variations in willingness to pay for preserving marble monuments: classic heterogeneity, random parameters, and mixture models. J Cult Econ 27:215–229

Morgan DKJ, Hicks BJ (2013) A metabolic theory of ecology applied to temperature and mass dependence of N and P excretion by common carp. Hydrobiologia 705:135–145

Morgan M, Ho Y (2018) Perception of Asian carp as a possible food source among Missouri anglers. Hum Dimens Wildl 23:1–8

Nishida M (1988) A new subspecies of the ayu, Plecoglossus altivelis,(Plecoglossidae) from the Ryukyu Islands. Jpn J Ichthyol 35:236–242

Nishizawa E, Kurokawa T, Yabe M (2006) Policies and resident's willingness to pay for restoring the ecosystem damaged by alien fish in Lake Biwa, Japan. Environ Sci Pol 9:448–456

Nunes PALD, Van Den Bergh JC (2004) Can people value protection against invasive marine species? Evidence from a joint TC–CV survey in the Netherlands. Environ Resour Econ 28:517–532

Pejchar L, Mooney HA (2009) Invasive species, ecosystem services and human well-being. Trends Ecol Evol 24:497–504

Pimentel D, Zuniga R, Morrison D (2005) Update on the environmental and economic costs associated with alien-invasive species in the United States. Ecol Econ 52:273–288

Polidoro BA, Carpenter KE, Collins L, Duke NC, Ellison AM, Ellison JC, Farnsworth EJ, Fernando ES, Kathiresan K, Koedam NE (2010) The loss of species: mangrove extinction risk and geographic areas of global concern. PLoS One 5:e10095

Revelt D, Train K (1998) Mixed logit with repeated choices: households' choices of appliance efficiency level. Rev Econ Stat 80:647–657

Roberts M, Cresswell W, Hanley N (2018) Prioritising invasive species control actions: evaluating effectiveness, costs, willingness to pay and social acceptance. Ecol Econ 152:1–8

Sala OE, Chapin FS, Armesto JJ, Berlow E, Bloomfield J, Dirzo R, Huber-Sanwald E, Huenneke LF, Jackson RB, Kinzig A (2000) Global biodiversity scenarios for the year 2100. Science 287:1770–1774

Scarpa R, Thiene M, Train K (2008) Utility in willingness to pay space: a tool to address confounding random scale effects in destination choice to the Alps. Am J Agric Econ 90:994–1010

Simberloff D, Parker IM, Windle PN (2005) Introduced species policy, management, and future research needs. Front Ecol Environ 3:12–20

Thresher R, Van De Kamp J, Campbell G, Grewe P, Canning M, Barney M, Bax NJ, Dunham R, Su B, Fulton W (2014) Sex-ratio-biasing constructs for the control of invasive lower vertebrates. Nat Biotechnol 32:424

Train K (2003) Discrete choice methods with simulation. Cambridge University Press, Cambridge. https://doi.org/10.1017/CBO9780511753930 pp.137-147

Train KE (2009) Discrete choice methods with simulation. Cambridge University Press, Cambridge

Trumbull DJ, Bonney R, Bascom D, Cabral A (2000) Thinking scientifically during participation in a citizen-science project. Sci Educ 84:265–275

Uchii K, Minamoto T, Honjo MN, Kawabata Zi (2014) Seasonal reactivation enables cyprinid herpesvirus 3 to persist in a wild host population. FEMS Microbiol Ecol 87:536–542

Wald DM, Nelson KA, Gawel AM, Rogers HS (2018) The role of trust in public attitudes toward invasive species management on Guam: a case study. J Environ Manag 229:133–144

Zastrow M (2018) Doubts raised over Australia's plan to release herpes to wipe out carp. Nature, https://doi.org/10.1038/d41586-018-02315-4

Chapter 4
The Use of Backcasting to Promote Urban Transformation to Sustainability: The Case of Toyama City, Japan

Kazumasu Aoki, Yusuke Kishita, Hidenori Nakamura, and Takuma Masuda

Abstract Envisioning urban sustainability demands to embrace divergent values of various stakeholders. Implementation of policies realizing city's future visions needs support from a wide range of general public. Hence, merits of participatory approach to backcasting scenario-making have been noted. Although experimenting such approach should be more encouraged, it remains to be seen whether lay citizens can generate their scenarios with required level of rationale and soundness. This chapter addresses this important, but yet unexplored concern by taking two potentially contrasting perspectives. One is "divergence" found in processes where citizens express their pluralistic interests and preferences in an unconstrained manner. The other is "convergence" found in where such a diversified plurality is circumscribed and composed to engender in outcomes some form of converged context. A trade-off relationship may arise between these two and the chapter seeks if any balance can be upheld. To explore such question, a participatory workshop was held in Toyama city, Japan where a handful numbers of citizens envisioned in two separate groups their desirable future via backcasting city's sustainable features. In analyses, outcomes of both groups' scenarios were compared and also index of consistency-based text structures endogenous to the scenarios was quantitatively gauged by computational simulation technique. Findings suggest that while a broad spectrum of socioeco-

This work was performed when the fourth author used to be a bachelor student at the University of Tokyo.

K. Aoki (✉)
University of Toyama, Toyama, Japan
e-mail: kzaoki@eco.u-toyama.ac.jp

Y. Kishita
The University of Tokyo, Tokyo, Japan

H. Nakamura
Toyama Prefectural University, Toyama, Japan

T. Masuda
MS&AD Systems Company, Tokyo, Japan

© The Author(s) 2020

45

O. Saito et al. (eds.), *Managing Socio-ecological Production Landscapes and Seascapes for Sustainable Communities in Asia*, Science for Sustainable Societies, https://doi.org/10.1007/978-981-15-1133-2_4

nomic and ecological elements was incorporated, they were yet founded upon a fairly good degree of logically coherent means-end based structures. The chapter then considers the meaning of such a balance for backcasting scenario-making with implications for further research agenda and future policy-making.

Keywords Urban transition · Future visions · Backcasting scenario · Participatory approach · Citizen dialogue · Sustainable society scenario (3S) simulator

4.1 Introduction

4.1.1 Background Issues of This Study

In recent years, it has become increasingly evident that various aspects of contemporary cities place too high a burden on the environment. Population, industry, commerce, energy, food consumption, and culture are some of the factors leading to rapid concentrations in urban areas in both developed and developing countries (Wolfram and Frantzeskaki 2016). Therefore, like Bulkeley and Betsill (2003) and others (Hodson and Marvin 2010; Loorbach et al. 2016; Hodson et al. 2017; Frantzeskaki et al. 2018) have stated, urban sustainability transitions have emerged as an urgent policy agenda concerning possibilities for making fundamental transformative changes on this current trend to keep cities from following unsustainable pathways.

In this regard, since the end of the 1990s, relevant disciplines (e.g., sustainability science, socio-engineering, sustainability transitions, and so forth) have argued the merits of envisioning cities' sustainable futures and contemplating possible pathways toward realizing such futures (Gallopin et al. 1997; Matsuoka et al. 2001; Glenn and The Future Groups International 2005). Such interests in scenario-making also extend to a wide range of issues including climate change, biodiversity, sustainable development goals (SDGs), and landscape approaches such as *Satoyama and Satoumi*, which promote socio-ecological production landscape and seascapes (Tress and Tress 2003; Carpenter et al. 2005; International Panel on Climate Change (IPCC) 2007; Ten Brink et al. 2010; Kanie 2017).

Among those scenario-makings, many now regard a design method (and also a way of thinking) called "backcasting" (BC), to be particularly promising (Robinson 1990; Dreborg 1996; Mander et al. 2008; Nishioka 2008; Kok et al. 2011; Kishita et al. 2016). BC scenario-making is defined and understood as a series of processes where stakeholders: (1) first craft visions of their ideal and desirable cities whose functioning standards and conditions are to be achieved in a relatively distant future (e.g., the year 2050 or beyond) and (2) think through pathways chronologically backward from the future to the current period in terms of what must be done to realize such visions.

Having received closer attention from both the academic and nonacademic, practical fields, however, the BC methods have the following shortcomings that have not yet been explored (Vergragt and Quist 2011; Kishita et al. 2017). (1) While BC scenario-making has mostly been carried out by relevant experts and professionals, involving lay persons in its process has not yet been researched to a great extent (Kishita et al. 2016; McLellan et al. 2017). Considering the fact that any processes relating to transitioning toward sustainable cities are direct and indirect consequences of collective decisions made by a wide array of stakeholders, nonexperts and nonprofessionals should also take initiatives in BC scenarios (Rotmans et al. 2000; Kasemir et al. 2003; Albert 2008; Umeda 2008; McLellan et al. 2017; Frantzeskaki et al. 2018). (2) The resulted BC scenarios are not yet formally or effectively implemented in practice and not having as much impact as already existing policies and measures. That is partly because it has been difficult to embed such participatory BC scenario-making into ongoing, conventional policy processes (Soria-Lara and Banister 2017; Kishita et al. 2016).

4.1.2 Analytical Perspectives and Research Questions

In order to overcome such shortcomings, researchers must take into consideration the fact that envisioning sustainability of future cities inevitably demands to embrace divergent interests, preferences, and knowledges of various stakeholders (Umeda 2008). That is not only because, as most often said in the field of sustainable science, the concept of sustainability itself reflects a bundle of multiple value systems (Kishita et al. 2010), but also because any cities have their own societal functioning that is of multifaceted configurations working in an interrelated, complementary manner (Hodson et al. 2017). In this regard, letting the general public participate in BC scenario-making certainly makes it more plausible to consider more diverged pluralistic opinions and relevant local knowledge to emerge and be expressed throughout the process (McLellan et al. 2017). After all, the citizens have an ultimate stake and say in the direction of their own future cities, and without ensuring their credible commitments and continuous cooperation in the longer term, there will be no effective endeavor toward urban sustainability transition. Also, the recent arguments on SDGs and biodiversity tend to view participatory approach as essential to empower grassroots citizens whose knowledge and experiences are key enabling elements for successful future visioning (Kanie 2017).

At the same time, however, citizen participatory processes inevitably increase the difficulties to mediate interests of individuals, especially when these interests are not bucked by rational reasoning or public-minded causes. In democratic nations, it has indeed been noted that citizens rather easily follow the majority, occasionally being swallowed up by a tide of enthusiasm. Note also that this type of skepticism, sometimes called "populism," has constantly been associated with more direct, mass participatory democracy. Though anecdotal, a Japanese local administrative official we once spoke with said: "Most of the time, citizens just demand what they want only for themselves."

We then must ask how such risk can be reduced and governed to be able to generate more participatory BC scenario-making with a satisfactory level of rational and soundness (van der Heijden 2005) and to mobilize such scenarios as a practical policy tool for developing cities' sustainable futures, which indicates a need for deep transformative change from the current orientation (Kishita et al. 2018). Therefore, in this chapter, we consider two potentially contrasting perspectives. One is "divergence" found as diverged processes and opinions where such lay individuals as general citizens act as dispersed dots, expressing their pluralistic interests, preferences, and knowledge in processes in an unconstrained manner. The other is "convergence" found as converged outcomes and structures where such a diversified plurality is circumscribed and composed as connected dots, engendering in the scenarios some form of converged context or bases normally found in practical policy documents and texts. Between these two extremes, one may find a trade-off relationship and this is the area on which we will focus to explore if any balance can be upheld.

In answering this question, one should note that resolving the trade-off is expected to be more difficult to achieve using the BC method, especially compared to the more conventional method called "forecasting" (FC). FC has long been the dominant practical method and is of incremental nature and thus more inclined to path-dependency. Under FC methods, policies and measures tend to be planned and made incrementally, allowing them to be based on and drawn from the current version. Thus, reliance on FC can more likely assure the generation of solutions which are compatible to and stable with the ongoing functioning and configurations of the societal system. Nevertheless, literature in sustainability science and sustainability transitions has noted that path-dependency must be overcome for a society to solve so-called "wicked problems" and be transformed into a more sustainable one. Problems are considered wicked in the sense that they only worsen if we take the ongoing functioning and configurations of the current societal system for granted and let solutions be based on and drawn from them (Robinson 1990; Zellner and Campbell 2015; Loorbach et al. 2016).

In contrast, BC is basically understood as methods that can more likely allow greater proactive and pathbreaking changes to be sought and accommodated through scenario-making. This means that the BC methods are more inclined to include divergent ways of thinking, and involve processes where more creative, innovative ideas and thoughts can emerge pluralistically and be expressed (Robinson 1990; Dreborg 1996). In line with this understanding, one study even argues that under the BC method, if it is for the sake of generating scenarios with more edgy, pathbreaking solutions, some degree of gap or incoherence that may be found in causal relationships between means and ends is not necessarily considered to be overly problematic (Kishita et al. 2016). Thus, we explored whether and to what extent such BC scenarios-making could be converged in a reasonably rationalized fashion via engendering some form of sound, coherent bases and structures being attached to the texts of the resulted scenarios (Albert 2008; Alcamo et al. 2008).

Below, the chapter articulates in the second section as to how we designed and implemented a series of workshops (WS) where lay citizens participated and lays out the methods and techniques used to ensure the participants' unconstrained, spontaneous dialogues while backcasting desirable future city. The third section then explains two verification approaches undertaken to explore the research questions drawn from the above-mentioned perspectives. The one corresponds to "divergence" concern to see if the participants' values were pluralistically expressed. The other corresponds to "convergence" concern to see if and to what extent coherence can be found. The fourth section provides analytical outputs and relevant discussions. In the final fifth section, we argue as to what the resulted findings mean to the use of backcasting scenario-making in policy practice and also point to the direction of further research and its significance.

4.2 Designs of Backcasting Scenario-Making: Citizen Participatory Workshop

4.2.1 Setting of the Workshop Held in the City of Toyama, Japan

4.2.1.1 Reasons for Choosing the Targeted City

We chose the city of Toyama in Japan to be the study area for BC scenario-making. Toyama city has 420,000 inhabitants. Its location is near the center of Honshu, the main island of Japan and is in the Hokuriku region, about 450 km northwest of the Tokyo Metropolitan area.

Today, Japan as a nation faces a series of severe and rapidly growing socioeconomic problems (e.g., sharp population decline, highly aging society with a declining birthrate, severe fiscal deficit, and excess concentration of population and industry in the Tokyo Metropolitan area) that are considered to be endemic to economically advanced nations. Among the other cities of a similar size in Japan, Toyama city is regarded to be proactively committed to dealing with such problems.

Since the 2000s, the local government of Toyama city (TCG) has implemented so-called compact city policies under which outside residents are induced and incentivized mostly with monetary subsidies to move to and live their lives in the city's center. Concentration of the city's population and societal functions in one particular geographic area enables TCG to diminish its administrative expenses.

In this line of policy deployment, improvement of the local public transportation network (made of light rail transit, railroads, and buses) has been vigorously sought by TCG to reduce citizens' heavy dependence on the daily use of private automobiles. Living without car use (meaning the city life within walking distance) has also been a long-term objective pursued over decades by TCG, serving to address two of the socioeconomic problems. One is climate change; switching from private automobiles to public transportations can cut CO_2 emissions. The other is social welfare; enhanced physical strength with better walking abilities and thus extended healthy life expectancy can cut medical expenses from rapidly skyrocketing for the elderly. In the eyes of TCG, all of these efforts enable building a city that is caring and friendly to both humans and the environment. As a result, Toyama city seeks to make itself the chosen destination among the other cities in Japan, with the hopes of attracting a larger population and increased investment (Toyama City 2017).

At the same time, however, TCG's compact city policies pose a thorny problem concerning how the city achieves greater levels of coordination and symbiosis between its central and periphery areas. In fact, we encountered a number of citizens who have raised their concerns and worries that there might be a potentially irreversible disparity occurring from prospective continuous decline in the administrative services and resources allocated to the peripheral, hilly, and mountainous area. Also, some of those tend to see that a series of socioeconomic revitalization measures—understood by TCG as one of the most pressing agenda and taken almost exclusively in the central area—has been determined by processes not being open enough to include a wider spectrum of individuals who can better represent grassroots and civil societal viewpoints. From these perspectives, it can be said that the city faces a governance issue where a way of making its collective decision should be attuned to a more direct citizen participatory initiative.

In addition, Toyama city has a unique urban setting where surroundings of the terrain make the world's rare landscape environment. Such a landscape endows the city with a 4000 m height difference which extends from the peak point of the Tateyama mountain range (3000 m height) in the south to the bottom of Toyama Bay (1000 m depth) in the north. TCG tries to take advantage of this particular geographical characteristic to appeal the city's attractiveness. That in turn means that, on the one hand, the city accelerates bolstering and upgrading of urban functions in its central area, but at the same time the city needs to conserve and enhance a wide array of natural capitals in its peripheral rural area. We thus argue that such a diverse socioeconomic and ecological landscape constitutes an interesting context in which how the participating citizens envision desirable features of their future city can be affected.

4.2.1.2   Membership of WS Participants

All the participants volunteered to participate with understanding of the purpose of the WS. The total number of such participants was 16, of whom 9 were male. They extended from teenagers to elderly people in their 70s. Everyone but two participants are Toyama residents. The two nonresidents have ongoing business interests in the city and travel into the city every week day. Thus, all participants are direct stakeholders in the path of the city's future. In 2016, the WS was held as a three-time event on August 6th and 27th and on October 22nd, each involving 5 to 6-h sessions in the afternoon with intermissions.

During the WS, the participants were evenly divided into two groups (Groups A and B) based on gender and age (see Table 4.1). Each group was facilitated by two of our authors, who took a content neutral stance, not intervening in the participants' talks in terms of substance. Also, the facilitators had prepared in advance the

Table 4.1 Membership of the Groups A and B	Group A	5 males (70s, 60s, 30s, 20s, 10s)
		3 females (40s, 40s, 20s)
	Group B	4 males (50s, 30s, 20s, 20s)
		4 females (70s, 50s, 40s, 20s)

documents describing the procedures and methodologies and applied them equally to the two groups' talks in order not to generate any managerial differences between groups. These were the measures being implemented for governing the WS because of our intent to achieve neutrality and comparability of the contents of both groups' scenarios. In this manner, we made it possible to analyze the different outcomes being generated from the same processes.

4.2.2 Methods and Techniques of Dialogue Among the Workshop Participants

4.2.2.1 Rules and Norms for Free Dialogue

All participants were asked not to reveal any of their attributes such as professions, positions, and titles throughout the entire period of the WS. We thought this rule was important because people tend to be intimidated or feel restrained while talking to someone who they feel holds a superior position or has more expertise. At the same time, the participants were guided "not to deny, dominate, and conclude" their dialogue in any way or at any time (Sakano 2011). At the WS, these rules and norms were applied to the two groups for the sake of creating and maintaining an arena in where each one of the participants can be spontaneous and independent with others and thus can think and express opinions and preferences freely.

In addition, by setting the scenarios targeting a relatively longer term (in this case the year 2064, 48 years from the time of convening the WS in 2016), we meant to craft visions as ones radically different from, and more creative than those of the status quo. Also, the year 2064 points to a more distant future if compared to targeted years of the TCG's policies on urban development and planning. Thus, we thought that the participants were incentivized to become liberated from and uninfluenced by the current policies being pursued by the TCG.

4.2.2.2 Use of Key Items for Guidance

At the WS, we introduced a list of "key items" in relation to the issues of the city's sustainability. Those key items were the ones actually used in the TCG's general administrative plan in effect from 2007 to 2016 (Toyama City 2007). The plan had used the list to provide examples of certain aspects concerning relevant policies and measures indicating how TCG makes its decisions on urban development and

Fig. 4.1 Alignment of
nodes and links of Logic
tree

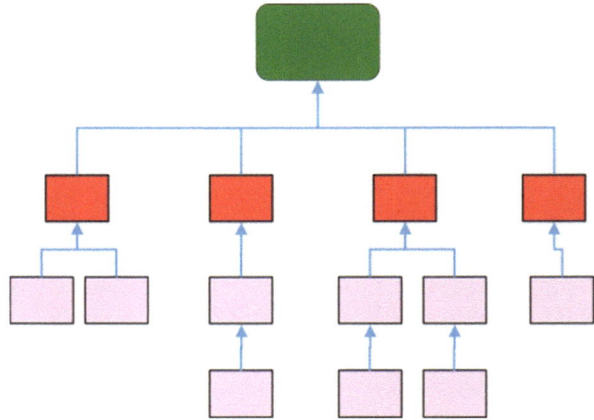

management. We also added to the list four broad elements: well-being, nature, economy, and society. These elements are equivalent to the ones proposed via a concept called a "sustainability compass" (AtKisson and Lee Hatcher 2001). Each broad element was divided into three subcategories respectively. For example, "well-being" consisted of "learning," "health," and "safety." Each subcategory then indicated a number of its own key items. For example, the key items listed for "learning" were "home education," "school education," and "life-long education."

During their dialogues, the participants were told to refer to the list while deliberating on the matter, but at the same time they were strongly advised not to be restrained by it as well. The reason for showing the list was founded on a procedural methodological perspective. It was meant to make sure that the multifaceted, pluralistic characteristics of urban sustainability issues were well contemplated by the participants and properly reflected the outcomes of the scenarios with the minimum necessary level and amount of dialogue.

4.2.2.3   Use of Logic Tree

As shown in Fig. 4.1, we employed an analytical instrument called a logic tree. Logic tree is a schematization tool used to visualize the dialogues' internal logics and underlying structures using a digraph method (Holcombe and Stein 1996; Wada et al. 2013). In its creation process, a top node represents a primary goal (green-colored node, top of Fig. 4.1) pursued in a vision of the scenario (e.g., a sustainable city). Under the primary goal, a series of secondary goals (red-colored nodes) (e.g., increased use of renewable energy, steady supply of resources and food, and extension of citizens' health expectancy) and means to achieve them (pink nodes) are put in a sequential manner so that a set of causal connections constituting the contents of the dialogues can be visually captured.

In accordance with the BC method, the WS participants were asked to set their goals first and then draw the means from such goals, not vice versa. In so doing, a chronologically backward way of thinking, a defining characteristic of BC, was maintained. During the WS, we the facilitators recorded the entire dialogues and used audio recordings as sources while constructing logic trees. Simultaneously, we also turned to documented outputs of the dialogues. Logic trees made after the first of the WS sessions were presented to the participants and utilized at the second and third WS sessions to induce and organize participants' ideas and thoughts. After the final WS (the third), we produced the final versions of the logic trees. Thus, the two groups' dialogues resulted in two logic trees.

4.2.2.4 Creating Multiple Scenarios

With regard to the concept of urban sustainability transition, we presupposed (as mentioned in Sect. 4.1) that when the participants' interests, preferences, and knowledge are expressed in a more pluralistically divergent fashion, it is more likely that some will be mutually exclusive and contestable. From our perspectives, it is vital that the scenario-making process reflects these as such because they too contribute to important, indispensable aspects of the concept of sustainability that is complex, inclusive, and comprehensive in its nature and because plurality of citizens' interests and preferences on urban sustainability should not be completely diminished in light of democratic collective decision-making (Kishita et al. 2018).

In that regard, we employed at the WS a method (Mizuno et al. 2012, 2013) that enables drawing multiple, not one, visions as a result of one set of dialogues. Such a method basically utilizes "key factors" that are prioritized by the participants in the process of crafting their visions. In more concrete terms, we turned to the method in which the participants first pick ten most important "key words" with reference to the output of the logic tree resulting from their dialogue. During this process, it was suggested that they refer to the list (see Sect 4.2.2.2), and again were advised not to be constrained by it when making their top ten choices.

They then selected the top two of the ten key words by scoring both the degree of importance and the degree of mismatch between the current actual conditions and an ideal state in the future. These top two are then named as "key factors." With regard to each of the top two key factors, the participants were led to discuss and define two contrasting, contestable conditions and functions of their desirability for a future sustainable city. By crossing these two key factors just like two axes bisecting each other at right angles, this method enables generating four different visions (see Fig. 4.2). Thus, at the WS, the two groups generated eight visions in total (four each, see also Fig. 4.4 below).

4.2.2.5 Choosing the Best Scenario

During the final phase, each group deliberated on which one to choose as the best from those four visions. It was important that they did not have to turn to a majority decision if they did not want to. Rather, they were advised by the facilitators to

Fig. 4.2 Crossing two key
factors that have two
contrasting features

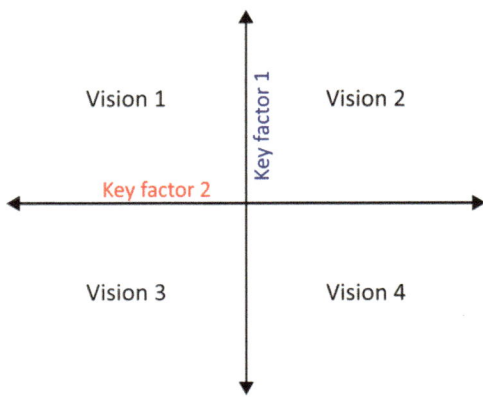

reach their consensus on how to decide their best vision to begin with. By employing this method, we sought to go beyond mere aggregation of established interests and preferences of the participants and to see if anything emerged to reshape and restructuralize a constellation of interests and preferences being manifested in the processes of their vision-making (Knight and Johnson 1994; Landemore 2013). The participants were to create a pathway only to the best vision, and a combination of such pathway and the best vision forms "the best scenario." This technique was used partly because of time constraints in the WS. As the final output, the WS resulted in the two best scenarios from the entire dialogues.

4.3 Verification Approaches to Research Questions

4.3.1 Examination of Multiple Visions

In verification processes, we first examined the processes and outcomes of the making of eight visions and the two best scenarios. In doing so, we basically look into an underlying question as to whether the contents of the visions could reflect and capture diverged interests, preferences, and knowledge of the participants despite the fact that the same methods and techniques (such as rules for dialogue, list of key items, logic tree, and scoring method) were evenly applied to the processes between Groups A and B. We in turn examine the processes by which the participants reached a consensus regarding the way to choose the best of their four visions, and in so doing, we also compare the outcomes of such visions between Groups A and B.

In taking these two approaches, we had presupposed that the more divergence reflected in the processes, the more the outcomes of the resulting four visions and best scenarios become independent of one another, meaning that they do not overlap in terms of their substance, by describing starkly different, or even mutually exclusive, states of the desirable future sustainable city.

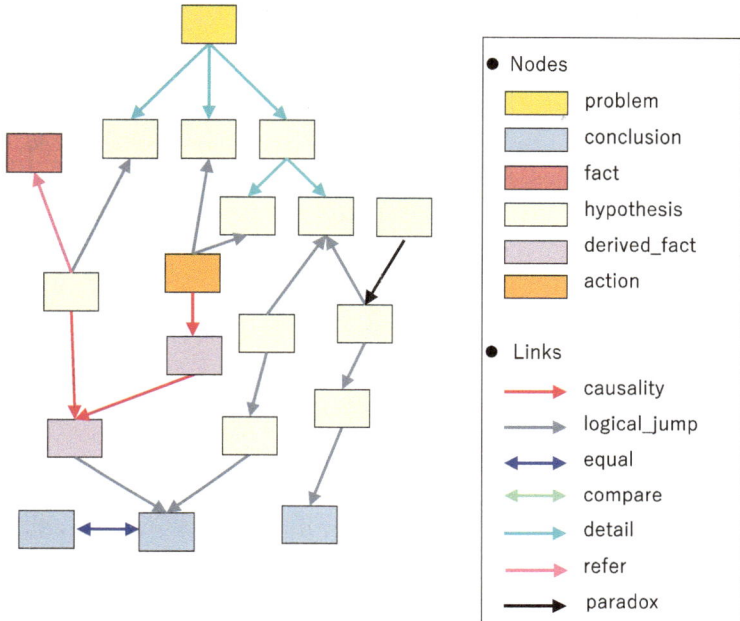

Fig. 4.3 Example digraph generated by 3S simulator

4.3.2   Use of the Sustainable Society Scenario (3S) Simulator

The second approach is the use of a simulation technique called the "sustainable society scenario simulator" (hereafter 3S simulator) proposed by one of the authors (Umeda et al. 2009; Kishita et al. 2009, 2010). The 3S simulator consists of an integrated supporting system developed for assisting us to comprehend, generate, and analyze scenarios concerning sustainable societies. The 3S simulator utilizes a computational model and algorithms to describe, with visualization, cause-and-effect relationships, logic, and structures that are endogenous to particular scenarios.

For instance, as shown in Fig. 4.3 below, importing outputs drawn from the logic tree analyses and from the audio recordings made for the WS sessions, the 3S simulator relies on the digraph method to visualize structures within the scenarios' contents by making linkages among a series of nodes. Each node and link are categorized in accordance with their attributes. Attributes of nodes that we used are: problem, conclusion, fact, hypothesis, derived fact, and action. Attributes of links are: causality, equal, logical jump, detail, refer, compare, and paradox (Umeda et al. 2009; Shelby et al. 2011). It can be understood that the order of these seven links basically indicates the degree to which each attribute consists of notions that are consistent with cause and effect and/or means-end relation. Thus, depending on how the attributes of nodes and links are aligned with one another, the degree of logical consistency attached to the contents of the scenarios can be evaluated. For instance, one can see that a scenario has a better logical consistency when its content can be described with a structure connected through a set of "causality" and "equal" links, rather than of "compare" and "paradox" links.

At the same time, however, it is important to understand that a "logical jump" link does not necessarily refer to a linkage that is logically inconsistent or false. It indicates a prospective inclusion of a breakthrough or innovative leap made within a boundary of limited "causality" that is yet to be of cause and effect and/or means-end relation. This loosened conceptualization of causality as a "logical jump" is required by the fundamental orientation of the BC method because the BC methodology is founded upon the rationale that a greater leap from the status quo and path dependency is needed to address the issues of sustainability transformation and its related structural changes (Kishita et al. 2009; Loorbach et al. 2016).

Furthermore, based on the above-mentioned methodology, we also turned to a numerical index system that quantifies the degree of such logical consistency. The index used is called the "logicality index" (hereafter LI) and is defined as the portion of arguments which are only constructed upon "causality" and "equal" links as compared to all other arguments. A formula used for our judgment hereby is that higher LI means better credibility of a scenario. In our concern for the efficacy and implementability of the BC scenarios as policy instruments, the degree of credibility can be an important indicator because it affects the extent to which BC scenarios can be accepted in society by embedding and incorporating them into the ongoing practical policy processes (Alcamo et al. 2008; Kishita et al. 2009).

In this study, therefore, we relied on LI to examine how credible the best scenarios generated by Groups A and B were. At the same time, we also turned to the 3S simulator to gauge LI in the existing scenarios of future sustainability. Note that all of such scenarios were made only by relevant professionals and experts and include ones that were made by following not only the BC but also FC methods and a combination of both. From our analytic perspective, whether or not the LIs of these future sustainability scenarios scored better than those of the best scenarios made during the WS, can be an important verification indicator.

4.4   Results, Analyses, and Discussion

4.4.1   Divergent Opinions Reflected in the Scenarios

Table 4.2 shows the lists of ten key words chosen by each group. One can see that: (1) notwithstanding the fact that many key words were chosen from the list of the "key items" (see Sect. 4.2.2.2) given to each group by the facilitators, the participants' came up with their own choices, (2) choices for the ten "key words" differ a great deal between the two groups.

In addition, as shown in Fig. 4.4, each group made very different choices on the key factors as well. One can also find that the important aspects the WS participants placed their visions are not similar to one another in the sense that each of the four visions has its own characteristics, and that these do not overlap in terms of direction and content.

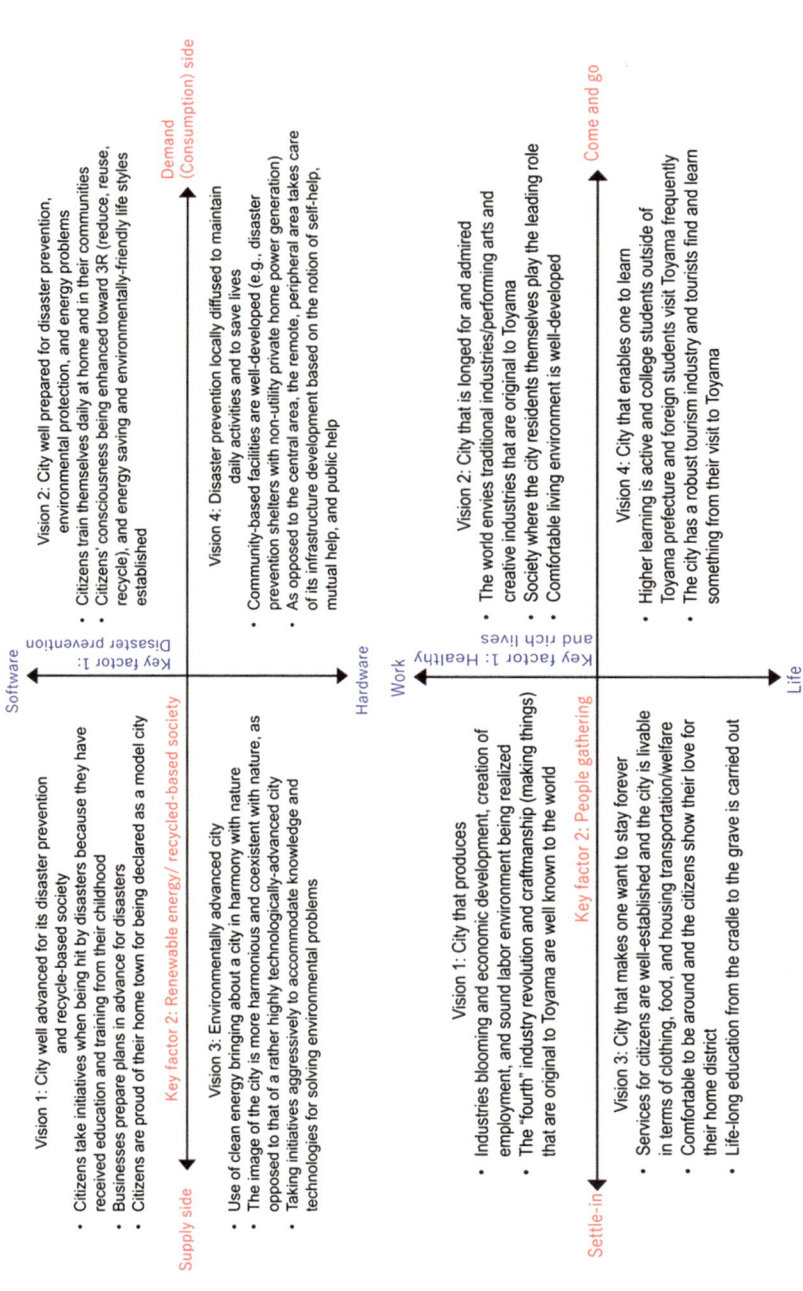

Fig. 4.4 Two key factors and four visions made by group A (upper) and B (lower) in comparison

Table 4.2 Differences in ten key words chosen by each group

Group A	Group B
Civic prides	People gathering
Disaster prevention	New culture
Health/welfare	Transportation
Renewable energy	Education
One of a kind community	Happiness
Internet of things	Good life/rich life/life of affluence
Way of living/way of working	Health
Tourism/migration (moving-in)	Vigorousness/bustling
Labor population	Harmony
Recycle-based society	Love for one's home town

Such diverged processes finally resulted in the two best scenarios whose visions' characteristics were significantly different from each other. The WS participants went on to discuss the paths that they thought necessary to realize such visions. Table 4.3 below shows the titles and some elements of the visions and paths. One can then be noted that the contents of these visions and paths reflected some important aspects of socioeconomic and socio-ecological concerns, which are elaborated as follows.

First, Group A's best scenario (Table 4.3, left column) indicated that considering the city's unique landscape characteristics (see Sect. 4.2.2.1), the WS participants proposed "dual residency" as a new type of lifestyle to achieve mutual harmony and benefit between the central and the peripheral areas. For this to be realized, they acknowledged that given the fact that prioritized allocation of TCG's budgets and resources would inevitably be held in the future, the citizens should turn off their ongoing tendency to depend on administrative services (i.e., government interventions and enterprises) and should instead have the mindset to cultivate and strengthen the spirit of self-help and/or mutual help compared to that of public help. They tended to regard such change in the division of roles between public and private spheres necessary to make their future city sustainable as a more unified entity.

Second, Group B's best scenario (Table 4.3, right column) contemplated that in order to earn people's admiration, the city would want capacity development in term of its production and education. That is because the WS participants thought that a place could attract people and have them stay there when residents felt enriched by and engaged in continuous, life-long learning that would also form the basis of sustainable production and provision of goods and services for its society. Thus, in the path toward 2064, implementation processes of creating Toyama's own industries and businesses ("city's original quaternary industry") were understood as vital. In such processes, the participants envisioned that the citizens should become as creative and unique as possible in their activities of daily living. They then acknowledged that such conduct could take place to a greater degree when the citizens fully learned and utilized the city's long-standing rich natural capitals (mountains, rivers, sea, paddies, and sceneries) and related cultural heritages (traditional arts and food cultures associated with water, rice, and fishes) that together constitute some important aspects of its socio-ecological landscape.

Table 4.3 Titles and elements of best scenarios (visions and pathways) in comparison

	Group A	Group B
Titles	Environmentally advanced city that is also harmonious and coexists with nature	A city that people around the world admire
Elements of visions	• Harmonious with nature for expansion of clean energy • The concept of dual residency between central and peripheral areas of the city is well received • As results of compact-city-related policies, public services and various facilities accumulate in the central area • In the peripheral area, residents solve local matters via cooperation under the spirit of self-help, mutual help, and public help	• To develop noble and distinctive humanity via vigorous and thorough life span inner education • To vitalize industries original in Toyama (traditional and/or entertainment industries, creative industries) • To introduce to and be admired by the world regarding citizens' distinctive humanity and the city's original quaternary industry • To establish an academic city and actively engage in conversational exchanges of ones' opinions
Elements of paths	[from 2016 to 2030] • To educate the citizens on disaster prevention and environment • To accumulate knowledges by inviting experts and making study tours to other cities • To designate model sections within the city [from 2030 to 2064] • To propose and accelerate the concept of dual residency between the central and peripheral areas • To develop clean energy that utilizes the resources of the Toyama prefectures	[from 2016 to 2030] • To secure fiscal revenues • To enhance inner education among citizens (moral education, performing arts, and discourses on religion and history) • To engage in public relations internationally [from 2030 to 2064] • To develop the city's own (original) industries via handing down of traditional arts and crafts and via innovating creative industries • To develop the city's infrastructures regarding scenery (landscape) and public transportation

4.4.2 Convergence Found from the BC Scenarios and Their Making

4.4.2.1 Merged into a Holistic, "Systemness" Perspective

It is important to note that the WS participants did not end up identifying mutually exclusive, contrasting features in their choices of "key factors." Described in Sect. 4.2.2.4 above, we had expected in advance that their divergent opinions can be better reflected by two different, possibly contrasting dimensions attached to the key factors. However, it appears that they saw mutually complementary features that together apparently constitute essential aspects of the city's societal functioning represented by each of the four key factors. For instance, one key factor "disaster prevention" is an essential aspect of the sound functioning of a societal system where a city seeks to become more sustainable in the future. Coming up with "software" and "hardware" to define and characterize such functioning is not only rightly attuned to divergence concerns, but is also attuned to bolster and leverage that functioning through utilizing some sort of holistic viewpoint or, we might rather say, a "systemness" perspective. This can be the case because disaster prevention can never properly function as a societal system without either of the two features. Also, it is more than significant that one can find such a systemness perspective running throughout the other three key factors as well.

Furthermore, we argue that this systemness perspective may also be found in the way in which both groups came up with their best visions. Each group's participants did not just pick one out of the four visions as the best. Instead, what each group did was to merge these four into a single vision. Group A, for instance, named such vision as "environmentally-advanced city that is also harmonious and coexists with nature" because, while centering on vision 3, its members also took elements and features (such as solving local matters via cooperation under the spirit of self-help, mutual help, and public help and being harmonious with nature for expansion of clean energy) from the three other visions. This took place simultaneously in both groups and occurred without their being instructed to do so by the facilitators and without their resorting to any types of majority decision.

As a consequence, we argue that their choosing the best vision apparently entailed a rather holistic, systemness perspective, by which the participants brought together aspects of societal functioning that they thought are needed for their version of the future desirable city to become more sustainable. If this is the case, we can point out that by finding a way to reach consensus on their own, the participants far exceeded our ex ante expectations in terms of going beyond the mere preference aggregation type of decision-making (see Sect. 4.2.2.5).

4.4.2.2 Textual Structures Backed by Logical Consistency

Based on the 3S analyses, we found that the contents of both groups' best scenarios were aligned with the form of structure visually expressed by a set of means-end chains linked among the nodes (Fig. 4.5). As mentioned in Sect. 4.3.2 above, this in turn means that both scenarios equally entailed a degree of internally coherent, logical consistency in each of their structures. Considering the fact that the two scenarios significantly differed in their substance (as discussed in Sect. 4.4.1), we claim that it is a worthy finding having some bearing on procedural legitimacy that might prove the efficacy and implementability of the citizen participatory approach and its processes toward BC scenario-making.

In terms of this internal logical consistency, what we also found from our 3S analyses was congruent with our claim. As shown in Table 4.4, when compared, the LI numbers for both best scenarios (19% for Group A, 10% for Group B) were as good as the ones scored for the BC scenarios being made by three groups consisting of experts only (20%, 15%, and 9%). This suggests that as far as the internal logical consistency attached to the scenarios' textual structure is concerned, the same degree of credibility and/or robustness can be acquired by either experts' or lay citizens' participation in BC decision-making.

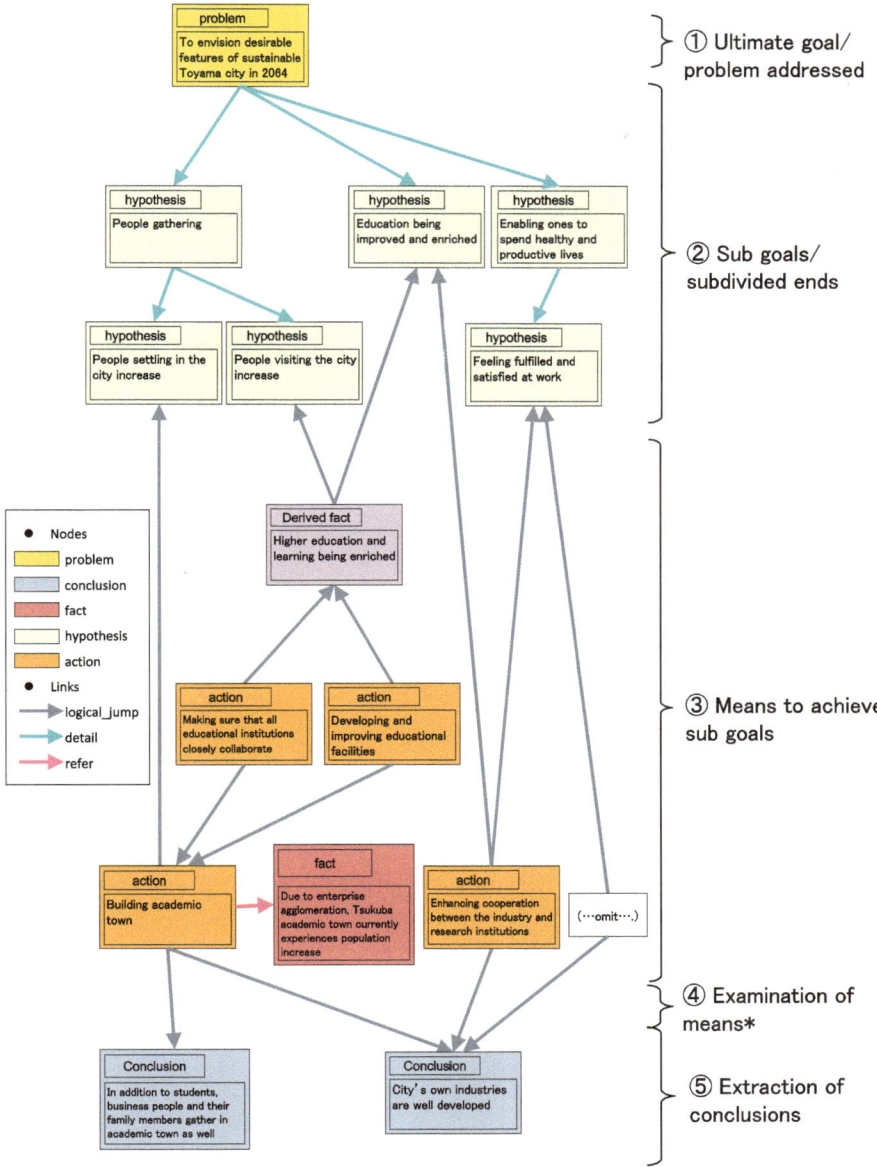

Fig. 4.5 Portion of the result of the 3S simulator analysis for Group B

Table 4.4 LIs of various scenarios in comparison

Scenarios	Methods	LI [%]
Group A's at Toyama WS	Backcast	19
Group B's at Toyama WS	Backcast	10
HEV diffusion scenario (Matsumoto et al. 2008)	Forecast	58
IPCC GHG emission scenario (IPCC 2007)	Forecast	47
ETP 2012 scenario (IEA 2012)	Forecast/Backcast	20
2050 Japanese low carbon society scenario (Nishioka 2008)	Backcast	15
Sustainable manufacturing industry scenario (Mizuno et al. 2014)	Backcast	9

In addition, the very methods and procedures we employed in the WS can receive further credit because while the LIs for the lay citizens' BC scenario-making produced similar scores in comparison with the experts, the LIs also were much lower than the ones scored under the FC method (58%, 47%). This in turn suggests that the intended functioning of the BC methods was secured as expected so that more logical jumps departing from the status quo and/or breaking through path dependency might be found in the BC rather than in the FC method.

4.4.2.3 Issue-Oriented Concerns Being Shared

Furthermore, our 3S analyses found another convergence, where both groups' best scenarios had the same set of the subgoals being incorporated into their means and ends structures (see Fig. 4.5, (2) subgoals/subdivided ends). Such subgoals are "civic pride/love for one's home town," "health/welfare," and "preventing declining population." The findings here indicate that the WS participants happened to share the same issue-oriented concerns, and those concerns are the ones that were actually considered and pursued through relevant policies and measures by the TCG. Also, this finding becomes more intriguing when we consider the fact that the ways in which the nodes were linked differed significantly between the two groups' scenario structures.

4.5 Conclusions and Implications for Further Research

Based on the analytical results shown above, we argue that while a broad spectrum of socioeconomic and ecological elements was incorporated, lay citizens' unrestrained dialogues in the WS sessions generated scenarios with a notably good degree of logically coherent, means-end-based structures and even projected what appeared to be a holistic, systemness-oriented perspective. In rhetorical terms, it can be said that once dispersed dots were connected to engender forms/texts backed by certain rational bases. Accordingly, we find it reasonable to conclude that when governed in certain ways, citizen participatory approaches can hit a fairly good balance between diverged processes and converged outcomes of BC scenario-making on the issue of urban sustainability transition. This leads us to claim that by way of ascertaining a required level of credibility, such a finding serves to prove and enhance practicality and implementability of BC scenarios that are considered daily policy instruments.

Among other possible contributing factors, the ones most significant to this favorable result would be the combination of logic trees and the list of key items. While it is plausible that these two aspects synergistically provided the WS participants with a good deal of reference focal points affecting the ways in which their dialogues were framed and constructed, each of the two groups ended up having their divergent opinions reflected in the scenarios' substances. Considering the fact that envisioning sustainability of future cities inevitably demands embracing pluralistic ideas and reflecting a set of multiple value systems (Umeda 2008; Kishita et al. 2010; Lang et al. 2012; Kanie 2017), we consider that the methods and processes we employed in the WS have some credibility toward achieving this goal.

Furthermore, our 3S analyses found that three issues—civic pride, health/welfare, and dealing with population decline—were equally recognized by both groups as important aspects to be pursued and realized when considering desirable features and functions of their future city. Even if those three issues were induced to be participants' significant interests because of the framing effects associated with the list of key items, we argue that this still conveys an important implication, particularly from a practical policy viewpoint. Notwithstanding the fact that the two groups came up with significantly different scenarios, it was shown that all the participants could reach a consensus on the importance and necessity of the same three policy-related issues. We tend to interpret such a phenomenon as manifestation of "core beliefs" held by the participating citizens acting as a collective decision-making entity. The significance of such a finding in BC scenario-making should not be undervalued because it is central to the democratic procedural legitimacy attached to collective decision-making (Sandker et al. 2010; Lang et al. 2012).

Drawing from such an understanding, we assert that further analyses should be conducted to identify what types of relationships and dynamics exist within the dialogues and processes between the agreed three issues (i.e., core beliefs in our terms) and the converged textural structures. When explored in a positivistic/empirical manner, findings thereof will make a significant contribution to deeper understanding of the interaction and its mechanism between the two concepts of "divergence" (diverged processes and opinions) and "convergence" (converged outcomes and structures) inherently attached to BC scenario-making, especially viewed in the context of general citizens' consensus-building. At the same time, such studies will also contribute to advancing the understanding of the complex and multifaceted nature of a future city's function and its transition toward sustainability.

We therefore claim that the results of our study indicate that there exists the rationale for bringing about more and more lay citizens' direct participation in BC scenario-making in the future. Besides being often advocated in the relevant literature from a normative standpoint, the theme has not so far been addressed or examined in terms of empirical research, analyses, or backing data. In such sense, this chapter is a pioneering effort to fill these gaps.

Acknowledgements We would like to thank Program for Promoting Regional Revitalization by Universities as Centers of Community (COC+ Program) sponsored by MEXT and JSPS KAKENHI Grant Number JP16K00671 for their support in making Toyama workshops and this chapter possible, as well as the two anonymous reviewers for their insightful comments. We would also like to thank citizen participants in the workshops for their proactive participation.

References

Albert C (2008) Participatory scenario development for supporting transitions towards sustainability. In: Proceedings of 2008 Berlin conference on the human dimensions of global environmental change "long-term policies: governing social-ecological change," 2008

Alcamo J, Kok K, Busch G, Priessm JA, Eickhout B, Rounsevell M, Rothman DS, Heistermann M (2008) Searching for the future of land: scenarios from the local to global scale. In: Land-use and land-cover change: local processes and global impacts. Springer, Berlin, pp 137–155

AtKisson A, Lee Hatcher R (2001) The compass index of sustainability: prototype for a comprehensive sustainability information system. JEAPM 3(4):509–532

Bulkeley H, Betsill M (2003) Cities and climate change: urban sustainability and global environmental governance. Routledge, London

Carpenter S, Pingali P, Bennett E, Zurek M (2005) Ecosystems and human well-being: scenario: findings of the scenarios working group, millennium ecosystem assessment series, vol 2. Island Press, Washington, DC

Dreborg KH (1996) Essense of backcasting. Futures 28(9):813–828

Frantzeskaki N, Hölscher K, Bach M, Avelino F (eds) (2018) Co-creating sustainable urban futures: a primer on applying transition management in cities. Springer, New York. https://doi.org/10.1007/978-3-319-69273-9

Gallopin G, Hammond A, Raskin P, Swart R (1997) Branch point: global scenarios and human choice. PoleStar Series Report, no. 7. Stockholm Environment Institute, Stockholm. https://greattransition.org/archives/other/Branch%20Points.pdf

Glenn JC, The Future Groups International (2005) 13. Scenarios. In: Glenn JC, Gordon TJ (eds) Futures research methodology version 2.0. AC/UNU Millenium Project, Washington, DC

van der Heijden K (2005) Scenarios: the art of strategic conversation. Wiley, Chichester

Hodson M, Marvin S (2010) Can cities shape socio-technical transitions and how would we know if they were? Res Policy 39:477–485

Hodson M, Geels FW, McMeekin A (2017) Reconfiguring urban sustainability transitions, analysing multiplicity. Sustainability 9(2):299. https://doi.org/10.3390/su9020299

Holcombe MW, Stein J (1996) Presentations for decision maker, 3rd edn. Wiley, Chichester

International Energy Agency (IEA) (2012) Energy technology perspectives 2012. IEA Publications, Paris

International Panel on Climate Change (IPCC) (2007) Climate change 2007: synthesis report. Contribution of working groups III to the fourth assessment report of IPCC

Kanie N (2017) What is SDGs?: agenda for transformation towards 2030. Minervashobo, Kyoto. (in Japanese)

Kasemir B, Jäger J, Jaeger CC, Gardner MT (2003) Public participation in sustainability science: a handbook. Cambridge University Press, Cambridge

Kishita Y, Yamasaki Y, Mizuno Y, Fukushige S, Umeda Y (2009) Development of sustainable society scenario simulator-structural scenario description and logical structure analysis. In: Proceedings of the 16th CIRP international conference on life cycle engineering, 2009, pp 361–366

Kishita Y, Mizuno Y, Fukushige S, Umeda Y (2010) Development of sustainable society scenario simulator - connecting scenarios with associated simulators. In: Proceedings of the 17th CIRP international conference on life cycle engineering 2010, pp 402–407

Kishita Y, Hara K, Uwasu M, Umeda Y (2016) Research needs and challenges faced in supporting scenario design in sustainability science: a literature review. Sustain Sci 11:331–347. https://doi.org/10.1007/s11625-015-0340-6.

Kishita Y, McLellan BC, Giurco D, Aoki K, Yoshizawa G, Handoh IC (2017) Designing backcasting scenarios for resilient energy futures. Technol Forecast Soc Chang 124:114–125

Kishita Y, Masuda T, Nakamura H, Aoki K (2018) Exploring a participatory approach towards designing backcasting scenarios in the city of Toyama, Japan: prospective visions and pathways to the city's sustainable future. J Econ Stud 64(1):127–152. (in Japanese)

Knight J, Johnson J (1994) Aggregation and deliberation: on the possibility of democratic legitimacy. Political Theory 22(2):277–296

Kok K, van Vliet M, Barlund I, Dubel A, Sendzimir J (2011) Combining participative backcasting and exploratory scenario development: experiences from the SCENES project. Technol Forecast Soc Chang 78:835–851

Landemore H (2013) Democratic reason: politics, collective intelligence, and the rule of the many. Princeton University Press, Princeton

Lang DJ, Wiek A, Bergmann M, Stauffacher M, Martens P, Moll P, Swilling M, Thomas CJ (2012) Transdisciplinary research in sustainability science: practice, principles, and challenges. Sustain Sci 7(suppl 1):25–43

Loorbach D, Wittmayer JM, Shiroyama H, Fujino J, Mizuguchi S (eds) (2016) Governance of urban sustainability transitions: European and Asian experiences. Springer, Japan. https://doi.org/10.1007/978-4-431-55426-4

Mander SL, Bows A, Anderson KL, Shackley S, Agnolicci P, Ekins P (2008) The Tyndall Decarbonisation scenarios-part I: development of a backcasting methodology with stakeholder participation. Energy Policy 36:3754–3763

Matsumoto, M., Kondoh S, Fujimoto J, Masui K (2008) A modeling framework for the diffusion of green technologies. In: Management of technology innovation and value creation selected papers from the 16th international conference on management of technology, 2008, pp 121–136

Matsuoka Y, Harasawa H, Takahashi K (2001) Scenario approach on global environmental problems. J Jpn Soc Civil Eng II-19(678):1–11. (in Japanese)

McLellan BC, Kishita Y, Aoki K (2017) Participatory design as a tool for effective sustainable energy transitions. In: Sustainability through innovation in product life cycle design. Springer, Singapore. https://doi.org/10.1007/978-981-10-0471-1_40

Mizuno Y, Kishita Y, Wada H, Kobayashi K, Fukushige S, Umeda Y (2012) Proposal of design support method of sustainability scenarios in backcasting manner. In: Proceedings of the ASME 2012 international design engineering technical conferences & computers and information in engineering conference: 17th design for manufacturing and the life cycle conference (DFMLC), DETC2012–70850

Mizuno Y, Kishita Y, Matsuhashi K, Miyake G, Murayama M, Umeda Y, Harasawa H (2013) An approach to designing sustainability scenarios part 1: a design method for backcasting scenarios. In: Proceedings of EcoDesign 2013, O-I-9

Mizuno Y, Kishita Y, Fukushige S, Umeda Y (2014) Envisioning sustainable manufacturing industries of Japan. Int J Autom Technol 8(5):634–643

Nishioka S (2008) Scenario for low carbon Society in Japan: a road to 70% reduction of CO_2 emission. Nikkan Kogyo Shimbun, Ltd (in Japanese)

Robinson JB (1990) Futures under glass: a recipe for people who hate to predict. Futures 22(9):820–842

Rotmans J, van Asselt M, Anastasi C, Greeuw S, Mellors J, Peters S, Rothman D, Rijkens N (2000) Visions for a sustainable Europe. Futures 32:809–831

Sakano T (2011) Experimenting a method of deliberation-based social survey called deliberative poll. In: Inohara T (ed) Consensus building: theory, methods, and practice. Keisoshobo, Tokyo, pp 141–159. (in Japanese)

Sandker M, Campbell BM, Ruiz-Perez M, Sayer JA, Cowling R, Kassa H, Knight AT (2010) The role of participatory modeling in landscape approaches to reconcile conservation and development. Ecol Soc 15(2):13

Shelby R, Perez Y, Agogino A (2011) Co-design methodology for the development of sustainable and renewable energy Systems for Underserved Communities: a case study with the Pinoleville Pomo nation. In: Proceedings of ASME 2011 international design engineering technical conferences and computers and information in engineering conference, 2011, pp 515–526

Soria-Lara JA, Banister D (2017) Participatory visioning in transport Backcasting studies: methodological lessons from Andalusia (Spain). J Transp Geogr 58:113–126

Ten Brink B, van der Esch S, Kram T, van Oorschot M, Alkemade JRM, Ahrens R, Bakkenes M, Bakkes JA, van den Berg M, Christensen V, Janse J, Jeuken M, Lucas P, Manders T, van Meijl H, Stehfest E, Tabeau A, van Vuuren D, Wilting H (2010) Rethinking global biodiversity strategies: exploring structural changes in production and consumption to reduce biodiversity loss. Netherlands Environmental Assessment Agency (PBL), Bilthoven

Toyama City (2017) Toyama City: people- and environment-friendly town (in Japanese). http://www.city.toyama.toyama.jp/special/eco.html. Accessed 8 Jan 2018

Toyama City, Department of Planning and Coordination (2007) General plan of Toyama City: 2007–2016 (in Japanese). http://www.city.toyama.toyama.jp/data/open/cnt/3/8180/1/zentai.pdf. Accessed 8 Jan 2018

Tress B, Tress G (2003) Scenario visualisation for participatory landscape planning: a study from Denmark. Landsc Urban Plan 64:161–178

Umeda Y (2008) The need and approach of sustainable society scenario simulator. In: Ecodesign 2008 proceedings of the Japan symposium, A12–1 (in Japanese)

Umeda Y, Nishiyama T, Yamasaki Y, Kishita Y, Fukushige S (2009) Proposal of sustainable society scenario simulator. CIRP J Manuf Sci Technol 1(4):272–278

Vergragt PJ, Quist J (2011) Backcasting for sustainability: introduction to the special issue. Technol Forecast Soc Chang 78(5):747–755

Wada H, Kishita Y, Mizuno Y, Fukushige S, Umeda Y (2013) Proposal of a design support method of Backcasting scenarios for sustainable society. Trans Jpn Soc Mech Eng C 79(799):845–857. (in Japanese)

Wolfram M, Frantzeskaki N (2016) Cities and systemic change for sustainability: prevailing epistemologies and an emerging research agenda. Sustainability 8(2):144. https://doi.org/10.3390/su8020144

Zellner M, Campbell SD (2015) Planning for deep-rooted problems: what can we learn from aligning complex systems and wicked problems? Plan Theory Pract 16:457–478

Chapter 5
Traditional Knowledge, Institutions and Human Sociality in Sustainable Use and Conservation of Biodiversity of the Sundarbans of Bangladesh

Rashed Al Mahmud Titumir, Tanjila Afrin, and Mohammad Saeed Islam

Abstract This chapter attempts to (a) identify the drivers of biodiversity degradation of the Sundarbans of Bangladesh, (b) present an alternative understanding on the measures for sustainable utilisation and conservation of resources and (c) suggest actions and policy alternatives to reverse the process of degradation and to move towards transformative harmonious human–nature interactions. While it is documented that the size of the Sundarbans of Bangladesh reduced and several floral and faunal species of the forest have been facing threat of extinction, the causes of continuous and unabated loss of the resources of this forest region have not been rigorously demonstrated. By challenging the mainstream approaches, the chapter theoretically and empirically exhibits that the exclusion of indigenous peoples and local communities (IPLCs) in the conservation and management process has contributed to the losses of biological diversity and suggests that the IPLCs have been practising several unique production methods based upon their traditional knowledge which can significantly contribute to the sustainable management of resources through symbiotic human–nature relationships. Following multiple evidence base (MEB) approaches, it is found that human sociality-based conservation practice positively impacts on resilient indicators and helps achieve Aichi Biodiversity Targets.

Keywords Human sociality · Traditional knowledge · Conservation · Biodiversity · The Sundarbans

R. A. M. Titumir (✉)
Department of Development Studies, University of Dhaka, Dhaka, Bangladesh

Unnayan Onneshan, Dhaka, Bangladesh
e-mail: rt@du.ac.bd; rtitumir@unnayan.org

T. Afrin · M. S. Islam
Unnayan Onneshan, Dhaka, Bangladesh

Department of Development Studies, Faculty of Arts and Social Sciences, Bangladesh University of Professionals, Dhaka, Bangladesh
e-mail: tanjila.afrin@bup.edu.bd; saeed.islam@bup.edu.bd

© The Author(s) 2020
O. Saito et al. (eds.), *Managing Socio-ecological Production Landscapes and Seascapes for Sustainable Communities in Asia*, Science for Sustainable Societies, https://doi.org/10.1007/978-981-15-1133-2_5

5.1 Introduction

The chapter considers the case of the Sundarbans of Bangladesh, the largest mangrove ecosystem of the world and a hotspot of biodiversity resources, to explore the underlying causes behind the continuous and unabated losses of its biodiversity resources and to seek viable means (policy) or measures (action) through which the process of degradation can be halted, the conservation process can be revitalised, and the sustainability of the resources can be ensured. It accordingly maps and finds out the key stakeholders and the agents dependent on the Sundarbans biodiversity resources and presents an alternative analysis to the sustainability of natural resources management integrating traditional knowledge (TK) systems to the socio-ecological production landscapes and seascapes (SEPLS) and draws on actions as regard to sustainable management of natural resources by means of harmonious human–nature nexus. Such an alternative analysis developed here can be used in other countries that are facing the same type of problems in biodiversity loss.

It is well documented that biodiversity resources have been declining at an alarming rate across different regions of the world posing threat to the future of humanity as well as to the other species (Higgins et al. 2013). Means and measures drawn on different school of thoughts are yet to find out the solutions of sustainable natural resource management which would lead to sustainable conservation process, secured livelihood options for the stakeholders and balanced ecosystem. Selfish resource exploitation, in fact, threatens societies as well as livelihoods contributing to a serious imbalance of the ecosystem (Battersby 2017). The situation is even worse in developing countries where the continuous pressures have already caused the extinction of numerous biodiversity resources. Bangladesh is no exception in this case. The Sundarbans of Bangladesh, known as the lung of the country, can now be identified as an important case of ecologically vulnerable area in terms of degradation of biodiversity resources. Several studies conducted on the Sundarbans have concluded that the resources of the Sundarbans have been declining gradually (e.g. Iftekhar and Islam 2004; Gopal and Chauhan 2006; Giri et al. 2007, 2014; Rahman et al. 2010; Rahman and Asaduzzaman 2010; Uddin et al. 2013; Islam 2014; Aziz and Paul 2015; Sarker et al. 2016). These studies have identified the external causes of forest degradation (e.g. conversion to other land use, over-harvesting, pollution, coastal erosion and climate change) or quantified the reduction in forest coverage area. Those studies, however, have not been able to provide solid theoretical foundation to analyse these problems and hardly propose an alternative suitable conservation and sustainability framework. Against this backdrop, this chapter critically explores the major theoretical underpinnings of neoclassical economics, institutional economics and political ecology to analyse the major drivers, including property rights instability, fragile institutions, lax regulatory regimes, unequal power sharing arrangements and political settlement. By employing such analyses of the state-of-the-art, the research exhibits that the exclusion of indigenous peoples and local communities (IPLCs) in the conservation and management process has contributed to the losses of biological diversity of the

Sundarbans. The chapter argues that the IPLCs have been practising several unique production methods based upon their TK which can significantly contribute to the sustainable management and conservation of natural resources through symbiotic human–nature relationships. It reveals, as a whole, that the well-being of SEPLS essentially depends on human sociality constructed by norms, values and other formal and informal institutions.

The next section presents a brief profile of the Sundarbans by identifying this mangrove ecosystem as a perfect case of SEPLS. The third section provides a conceptual framework that helps identify the major drivers of biodiversity resource degradation of the Sundarbans as well as examine the alternative means and measures for the conservation and sustainable utilisation of those resources. In the analyses parts of sections four and five, the empirical evidences have been discussed by juxtaposing the existing policy and institutional set up into the developed conceptual framework to reveal the major drivers of resource degradation and show alternative options which can be applied as viable means to manage the resources in a sustainable way. The penultimate section discusses the current resilience capacity of the Sundarbans based on the major findings of the study. The final section ends with concluding remarks.

5.2 A Brief Profile of the Sundarbans: A Socio-Ecological Production Landscape and Seascape (SEPLS)

This chapter uses three elements, here, in the form of *structure, benefits and changes* (Ichikawa, 2013) to present the Sundarbans as a perfect case of SEPLS.

5.2.1 Structure: Dynamic Mosaics of Habitats and Land Uses

The Sundarbans is located at the great delta of the Ganges, Brahmaputra and Meghna (GBM) rivers at the edge of Bay of Bengal and is the largest contiguous single-tract mangrove ecosystem in the world (Fig. 5.1). This mangrove ecosystem lies within both India (the State of West Bengal) and Bangladesh. The Bangladesh part is larger compared to the portion in India, with an area of 6071 km^2 (62% of the total area), which constitutes 39.5% of the total forest area of Bangladesh (Roy and Alam 2012). Of this Bangladesh part, 70% is land area and the rest (30%) is water (Kabir and Hossain 2008). The wetlands of the Sundarbans consist of about 200 islands separated by about 400 interconnected tidal rivers, creeks and canals (Rahman et al. 2010). The Sundarbans was recognised as a Natural World Heritage Site in 1997 by UNESCO and as a Ramsar Site of international importance in 1992 (IUCN Bangladesh 2014).

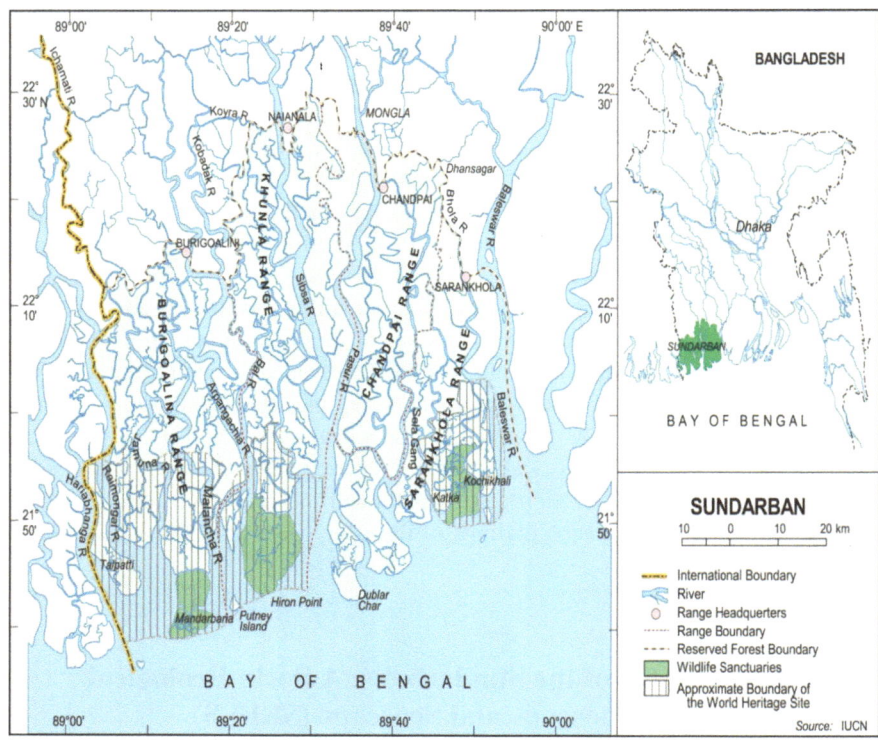

Fig. 5.1 Location of the Sundarbans. (Source: IUCN n.d as cited in Rahman et al. 2010)

5.2.2 Benefits: Maintaining Biodiversity and Providing Humans with Goods and Services

The Sundarbans harbours 334 species of trees, shrubs, herbs and epiphytes and about 400 species of wild animals (Behera and Haider 2012). Sundri (*Heritiera fomes*) is the most important floral species. Other prominent species are: gewa (*Excoecaria agallocha*), baen (*Avicennia officinalis*), passur (*Xylocarpus mekongensis*), keora (*Sonneratia apetala*), goran (*Ceriops decandra*), ora (*S. caseolaris*) and hental (*Phoenix paludosa*). It also offers high value non-timber forest products like honey, wax, fish and crabs. This forest is also rich in its faunal diversity. There are 448 species of vertebrates including 10 amphibians, 58 reptiles, 339 birds and 41 mammals (Department of Environment [DoE], Government of Bangladesh [GoB] 2015). It provides habitat for diverse aquatic wildlife such as estuarine crocodile (*Crocodylus porosus*), turtles (*Lepidochelys olivacea*), dolphins (*Platanista gangetica* and *Peponocephala electra*) and molluscs like the giant oyster (*Crassostrea gigas*). Nevertheless, the Royal Bengal Tiger (*Panthera tigris*) is the most magnificent animal. According to the census of 2004, around 440 tigers resided in the Bangladesh part while the most recent estimate puts such to around 106 tigers

(Bangladesh Forest Department [BFD], 2015 and The Guardian, 27 July 2015). It is also home to thousands of spotted deer (*Axis axis*) and barking Deer (*Muntiacus muntjak*).

These *biotic* along with other *abiotic* resources of the Sundarbans contribute directly or indirectly to the economy both at local and national levels. Fig. 5.2 shows how the resources of the Sundarbans have been utilised for different purposes, contributing both to the lives and livelihoods of local people and to the economy of the country. The livelihood pattern in the Sundarbans area varies with seasons and supports an estimated 3.5 million people directly or indirectly (Sarker et al. 2016). Wood and *golpata* collectors (*Bawalis*), fisherman (*Jele*), honey and wax collectors (*Mouals*), shell collectors (*Chunary*) and crab collectors are among the major occupational groups of the adjacent forest region. The lives and livelihoods of the local people are mainly related to the physical and biological (or biodiversity) resources as depicted in Fig. 5.2.

5.2.3 Changes: Shaped by the Interactions Between People and Nature

The Sundarbans has experienced major ecological and physiographical changes due to anthropogenic pressures and climatic disorder, which have taken a heavy toll on the regenerative capacities of the forest and its ability to maintain sustainability. Such pressures have resulted in the continuous decline of the forest coverage and of its biodiversity resources. In 1776, the size of the Sundarbans was 17,000 km^2. At present, it is only almost half of this total area (Islam and Gnauck 2009). A recent report shows declining trends in forest areas both in India and Bangladesh (Fig. 5.3).

The reduction of volume of important tree species of the Sundarbans can also be analysed through forest inventories prepared by different agencies (Table 5.1). The trend in growth of trees in each case is found to be declining.

The degradation of floral diversity also yields negative impacts on faunal diversity. As many as 20 globally threatened species inhabit in the Sundarbans. The most endangered species are *Batagur baska* (turtle), Ganges River dolphin and the Irrawaddy dolphin. Other threatened wildlife species include pythons, king cobras, adjutant storks, white-bellied sea eagles, clawless otters, masked fin-foots, ring lizards, river terrapins, fishing cats, spoon-billed sandpipers, and eagles (Department of Environment [DoE], GoB 2015). The most important faunal species, the Royal Bengal Tiger, is also enlisted as an endangered species by the IUCN. Table 5.2 provides a summary of the characteristics of the Sundarbans as regards SEPLS.[1]

---

[1] The two major indicators for identifying SEPLS have been specified here based on the definition by the *Satoyama Initiative* and illustrated by others (e.g. Gu and Subramanian 2012; Ichikawa 2013; Bergamini et al. 2013).

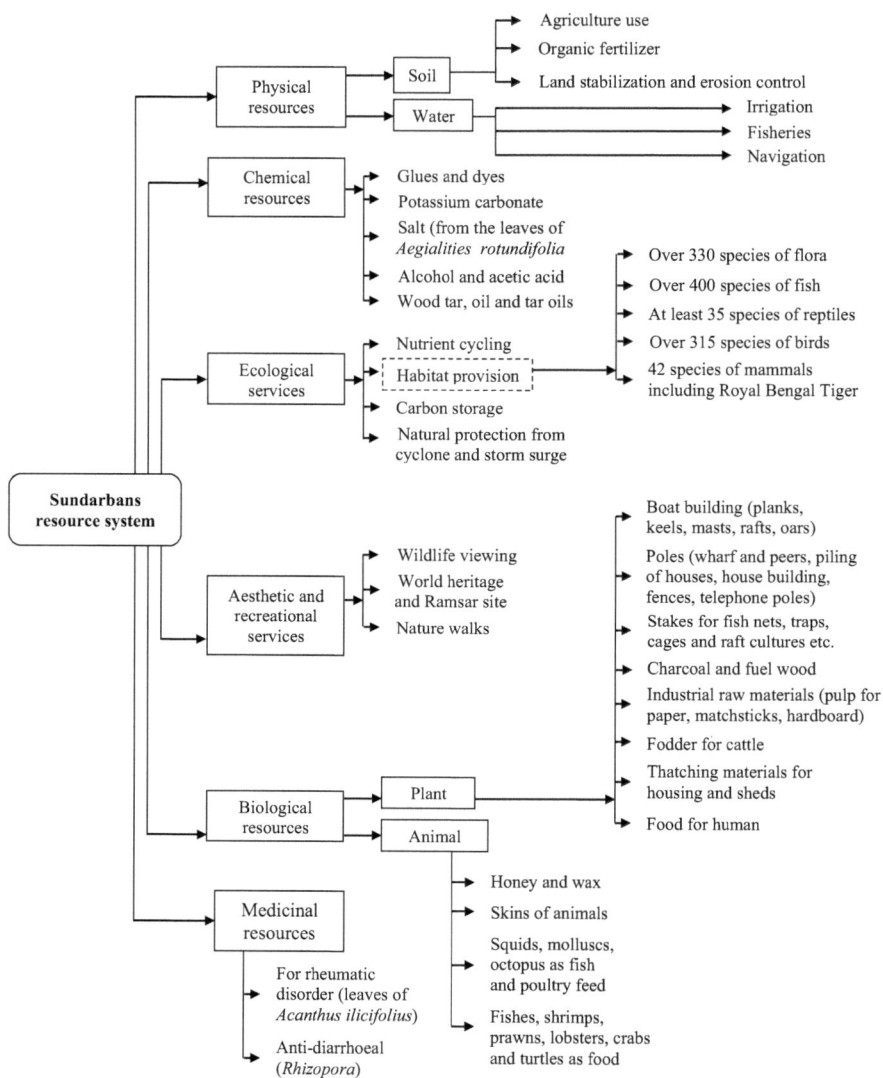

Fig. 5.2 Sundarbans resource system. (Source: Titumir and Afrin 2017)

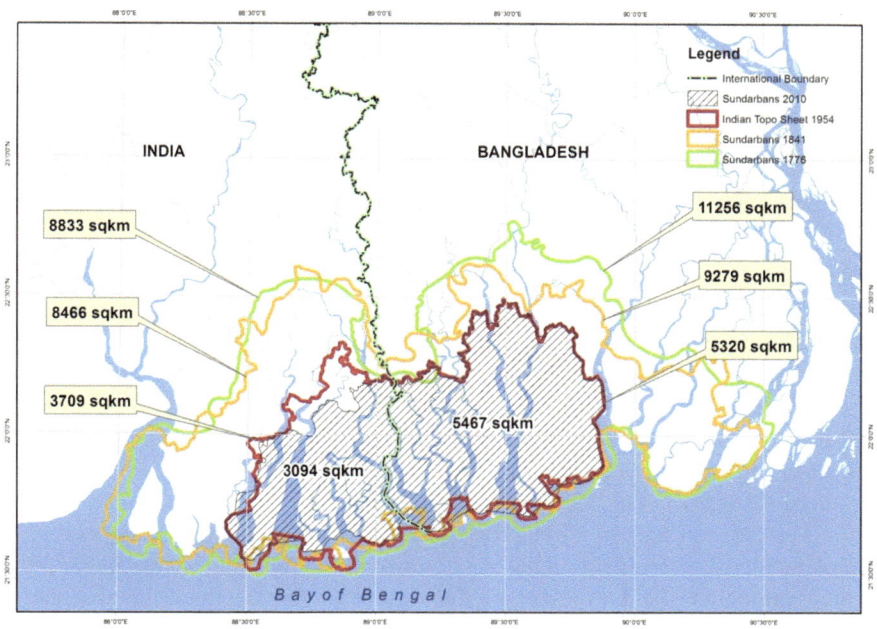

Fig. 5.3 Mangrove forest change of the Sundarbans from 1776 to 2010. (Source: Joint Landscape Narrative by India and Bangladesh, CEGIS 2016)

Table 5.1 Growing stock of the Sundarbans (Source: FAO 2011)

Year	Inventory done by	Sundri (number of trees per hectare)	Gewa (number of trees per hectare)	All tree species (number of trees per hectare)
1959	Forest and Forestal Engineering, Canada	211	61	296
1983	Overseas Development Authority	125	35	180
1996	Forest Resource Management Project, FD, GoB	106	20	144

Table 5.2 The Sundarbans as a SEPLS (Source: Titumir and Afrin 2017)

Indicators	Relevant to the Sundarbans? (yes/no)	Why relevant?
Mosaic of production landscape/seascape	Yes	It is a mangrove forest that includes forest, coastal and wetland ecosystems, supporting diverse production activities
Harmonious interaction between humans and nature and well-being of both	Yes	It provides the IPLCs different options for maintaining livelihoods and the IPLCs provide protection to the forest and its resources through traditional livelihood practices

It should, however, also be noted that the balance of such a SEPLS has continuously been threatened as has been found in the above discussion.

5.3   A Conceptual Framework: SEPLS, Human Sociality and Sustainability

Means and measures employed for the natural resource management are primarily drawn from market centric theoretical underpinning as a part of the intellective project of neo-liberalism. This school of thought suggests that the biodiversity resources degrade primarily because of the non-existence of market and negative externality (Sadmo 2015; Perrings et al. 1992). It argues that valuation techniques can provide useful insights to support policy initiatives by quantifying the economic value of the resources and to devise exchange rule associated with the protection of biological resources (Costanza et al. 1997; Pearce 2001; Bräuer 2003; Kumar 2005; Barbier 2007; McAfee and Shapiro 2010; Hahn et al. 2015). This understanding has been complemented by the institutional economists as establishing a formal property rights regime can efficiently manage the natural resources where the absence of property rights results in resources degradation (Ostrom 2000; Vatn 2009, 2010; Ituarte-Lima et al. 2014).

A section of the political economy analyses, on the contrary, contend that the existence of overlapping property rights regime contributes to the conflicting resources management and degradation. It sheds light on the political elements in resources management regime and highlights the hierarchical relationship that exists in society. It argues that institutional arrangements (property rights) are vulnerable to some political economic factors stemming from accumulation by different agents in presence of non-cooperative solution. It further stresses upon the roles of the formal political institutions and emphasises on the narratives about the changes of the ecosystem services (Robbins 2012).

Such literature provides a lens to describe the bio-environmental relationship in the presence of distribution of power to production activities and its link to ecological analysis (Greenberg and Park 1994). It emphasises on the claim that the degradation of natural resources is not only about the non-existence of market but also about unequal power sharing by the stakeholders over the management of resources (Fig. 5.4). Existence of vertical relations in society and upward enforcement of rules

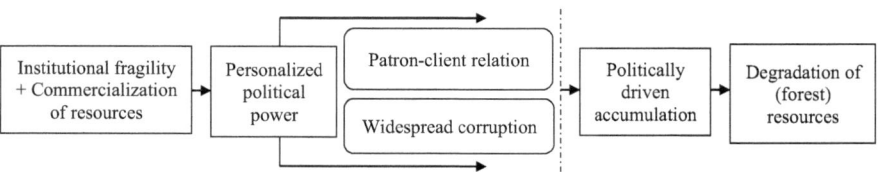

Fig. 5.4 Political economy factors inducing biodiversity resources degradation. (Source: prepared by the authors)

enable the powerful group to capture resources with impunity (Adhikari and Goldey 2010). The process prioritises the rule of individuals over the rule of law which ultimately results in institutional fragility, enlarging rent dissipation, rent seeking and seize of property rights.

Market centric analysis does not recognise that if particular species of ecosystem of a special kind are being traded for monetary gain, they might not be replaced. It, however, fails to offer a sustainable solution regarding the distinct characteristics of interdependent relationship among humans, biodiversity resources and ecosystems services. Exchange based on economic valuation is found to be faulty (Kosoy and Corbera 2010; Gomez-Baggetthun and Ruiz-Perez 2011; Muradian et al. 2013; Turnhout et al. 2013; Neuteleers and Engelen 2015). It reduces biodiversity into a number of quantifiable parts, subjecting to the utilitarian usage and reducing social–natural relations to market transactions (Turnhout et al. 2013). Such measures provide a narrow conception of ecosystem services and are potentially detrimental to the conservation of resources. Alongside, the political ecology does not provide any measures but a broad understanding of the contributing elements of the degradation of natural resources.

Human beings are part of the ecology not merely the exclusive agents who extract resources. The long-standing embeddedness of the human beings into the ecology and the roles they play into the system remains unexplored and sometimes has been identified as external to the system. Being a part of this system, human beings have been maintaining an interwoven, intimate and reciprocal nexus with the nature. This nexus can be explored from 'human sociality' perspective. Human sociality refers to the human beings, as a collective organisation, and is part of the larger ecosystem, which possess distinct knowledge and practices that systematically and sustainably contributes to the conservation and regeneration of the resources along with maintaining provision of ecosystem services. It stresses upon that societies in harmony with nature contribute to the biodiversity conservation through revitalisation and supporting SEPLS where informal institution plays a crucial role. Informal institutions which include norms, values and traditional knowledge not only contribute to the SEPLS but also conserve and regenerate the resources for making a more resilient ecological system and society.

A sustainability conservation framework constructed in this chapter exhibits that inter-institutional pitfall stemming from exclusion of informal institutions and community ownership causes degradation of the natural resources, contrary to the market-centric perspectives. It argues that the earlier practices of fencing off pieces of nature as a means to 'mitigate' anthropogenic intervention have been proved costly, unsustainable, and dubious in terms of socioeconomic and conservation processes (Liu et al. 2012). This alternative framework has taken the political economy premise to identify the causes of degradation with emphasising on the complementary relations between human beings and nature in ensuring the sustainable utilisation and distribution of the resources. It claims that conservation requires acknowledging a diversity of values, knowledge and framings of SEPLS which build the cooperation and incentivise conservation for long-term sustainable use of those resources (Fig. 5.5).

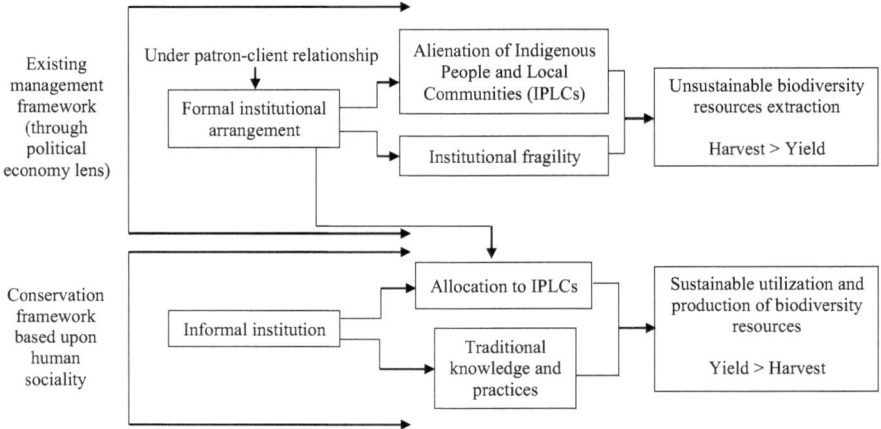

Fig. 5.5 Conceptual framework. (Source: prepared by the authors)

This framework argues that in the presence of neo-liberal means and measures, the exchange process constitutes a patron–client relationship. In this process, the IPLCs become the external agents to the ecological milieu, and it brings institutional fragility because of unequal power sharing between political elites and IPLCs. Such exchange relationship culminates into primitive accumulation of the resources and unsustainable extraction of resources (where, harvest is greater than the yield due to maximum realisation of the resources rent). Alternatively, the sustainable conservation framework based upon human sociality suggests that allocation of resources regime to the IPLCs is sustainable. IPLCs together with their traditional knowledge and practices constitute a socio-ecological production network. IPLCs contribute to sustain this production network because of its symbiotic nature to the stock of resources. This incentivises IPLCs to invent knowledge to conserve the resources and to practice the knowledge for ensuring a sustainable value chain. Thus, altogether the IPLCs and their TK practices make the biodiversity resources more resilient (where yield is greater than harvesting) and sustainable.

5.4 Drivers of Biodiversity Resource Degradation of the Sundarbans[2]

It is necessary to define the nature of property rights of a particular type of resources in order to identify the drivers of degradation of those resources through the lens of political economy. The reason is that fragile institutional arrangement (e.g. instable property rights) is at the root of resource degradation which results from the influ-

---

[2]The empirical sections (Sects. 5.4 and 5.5) discuss these results drawn from different studies, conducted by the *Unnayan Onneshan*(e.g. Kabir and Hossain 2008; Baten and Kumar 2010; UO 2010; Titumir 2011, 2015; Titumir and Afrin 2017; Titumir et al. [in progress]).

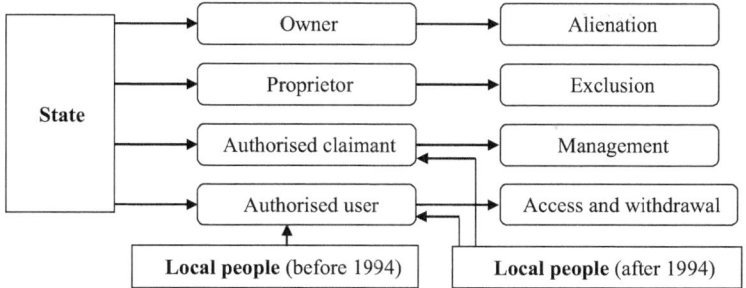

Fig. 5.6 Property rights structure of the Sundarbans. (Source: prepared by the authors)

ence of many political economic factors. A brief overview has been provided here firstly on the current structure of property rights of the Sundarbans. The major drivers of resource degradation have been identified thereafter.

5.4.1 Structure of Property Rights of the Sundarbans

The nature of property rights of the Sundarbans was ambiguous since formulation process. It was treated as open access forest for harvesting and conversion for agriculture particularly during Mughal period. The British colonisers ruling over Indian subcontinent became aware of the importance of this mangrove forest and declared it as Reserve Forest (RF) in 1878. The right over the forest was, thus, kept in the hands of the government. After the independence of Bangladesh in 1971, the forest of Bangladesh part was declared as RF again under the Forest Act 1927. Then, the Forest Policy of 1994, however, recognised the community participation in the management process and accordingly recognised the rights of the local people. The property rights structure of the Sundarbans now, therefore, cannot be defined in terms of specific type of property rights (common or public) rather the rights are being distributed among the state authority and local people. The overall structure of property rights can be explained through a diagrammatic representation based on Schlager and Ostrom's (1992) typology of bundle of property rights (Fig. 5.6).

Since 1994 the Forest Department (FD) on behalf of the state took the responsibility to ensure the efficient use of resources of the Sundarbans as the owner, proprietor, authorised claimant and authorised users. The resource users have the right to access and use resources by obtaining permission from the FD. On the contrary, the local people had got management rights along with the access and withdrawal rights. The practical scenario, however, signifies that this formal institutional arrangement is not stable. They have to face many barriers to exercise their rights to have access inside the forest and to use the biodiversity resources. Moreover, the FD is also found to be inefficient to exercise its legal rights in a stable way. Such instability is apparent through several legal and quasi-legal interventions by different powerful agents into this resourceful region as will be clarified in the below discussions.

5.4.2 Increasing Habitation and Illegal Encroachment

The existence of instable and ill-defined property rights creates scope for the politically and economically powerful groups to encroach into the forest of the Sundarbans in illegal ways. The Sundarbans, particularly, locates within the three districts of Khulna, Satkhira and Bagerhat. The density of settlement across these three regions has been increasing over the years, and the trend will continue as the projection indicates (Fig. 5.7). Shear dependence on natural resources of the Sundarbans, therefore, is also increasing. Such increasing habitation is largely an outcome of fragile property rights regime by the community over this ecological landscape. A significant number of migrated people find it possible to encroach into the forest and, therefore, intend to live in the nearby districts of the Sundarbans.

They are not the indigenous local people, and therefore, they do not respect the local customary practices to conserve the forest resources and always intend to extract the resources as much as possible and thus enhances the process of degradation. Moreover, politically and economically powerful groups are also found to continuously encroach into the forest region by making coalition at different levels.

5.4.3 Rent-Seeking Tendency and Extra-Legal Management

The government agencies, officials and functionaries are alleged to be rapacious in their own right too. There are irregularities in fishing and collection of honey, timber and *golpata (Nypa fruticans)*. For instance, in every case the traditional collectors have to get access right (BLC—Boat License Certificate) from FD to enter into the forest by paying extra tolls in form of bribe. To cope with such excessive tolls, the

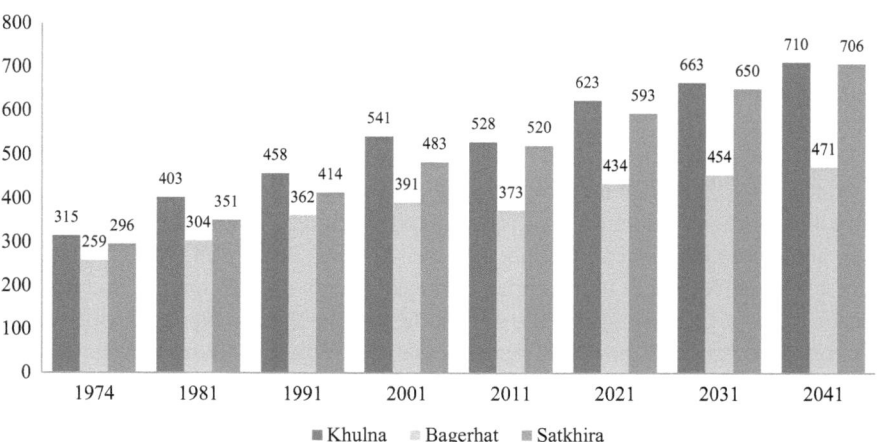

Fig. 5.7 Population density in the districts encompassing the Sundarbans (in number). (Source: Authors' calculation based on population census of 2001 and 2011 by BBS (2011))

resource collectors have to collect resources more than they are permitted to which adversely affects the reproduction capacity of the forests. Moreover, the illegal encroachment into the forest, as described in the previous subsection, by the politically powerful ones has been possible with the direct cooperation of forest officials through bribery and other illegal means such as embezzlement and misuse of power. Going against its own policy, the government over the last few years permitted setting up of 190 industrial and commercial units in the ecologically critical area (ECA) of the Sundarbans, which poses a serious threat to the biodiversity (Fig. 5.8). The government declared the 10-km periphery of the mangrove forest as the ECA in 1999, after the UNESCO listed it as a natural world heritage site. As per Bangladesh Environment Conservation Act 1995 (amended in 2010), no one is allowed to set up any factory in the ECA.

Most of these agents and interest groups of land grabbers are businessmen and industrials units who have powerful political linkage. The most recent and controversial project is the 'Rampal Power Plant Project', a coal-based power plant, fraught with triple jeopardises in the three domains of environment, economic and technical feasibility, which may cause dangers to the integrity of the Sundarbans. The project is under the process of implementation.

5.4.4 Land Reclamation and Shrimp Cultivation

Conversion of land into commercial shrimp farming is the largest human threat to the Sundarbans mangrove ecosystem. The increase of the farms is mainly caused through quasi-legal intervention. The farms are put in place by the powerful local stakeholders, specifically, by the rich fishermen (not part of the indigenous people), connected with political and administrative structures at local and national levels.

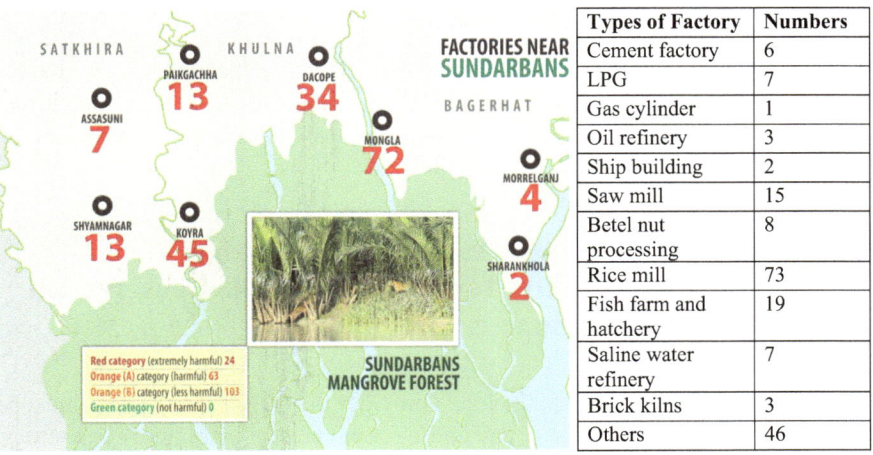

Types of Factory	Numbers
Cement factory	6
LPG	7
Gas cylinder	1
Oil refinery	3
Ship building	2
Saw mill	15
Betel nut processing	8
Rice mill	73
Fish farm and hatchery	19
Saline water refinery	7
Brick kilns	3
Others	46

Fig. 5.8 Factories near the Sundarbans. (Source: The Daily Star, 6 April 2018)

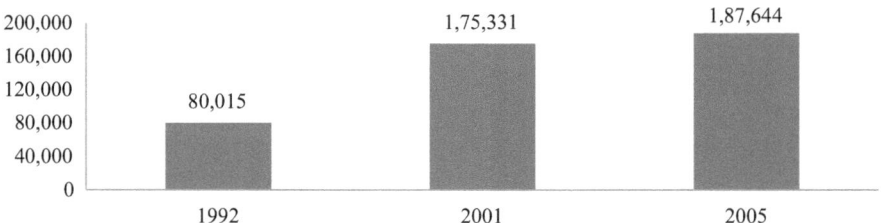

Fig. 5.9 Bagda shrimp cultivated areas adjacent to the Sundarbans (in hectares). (Data Source: Hussain 2014)

There is an increasing trend of shrimp (Bagda, *Penaeus monodon*) cultivated areas adjacent to the Sundarbans (in hectares) from 1992 to 2005 (Fig. 5.9). The constructions of shrimp ponds contribute to the degradation and loss of mangrove habitats in several ways. For instance, a shrimp-cultivating pond exhausts its usefulness within 3–6 years of construction.

Therefore, the cultivators have to move along the coast, destroying mangroves to make room for more ponds. Moreover, it increases salinity in the soil and thus alters the soil composition of that region. Southwest coastal region of Bangladesh is already facing increasing salinisation, especially between October and May. Laboratory analyses of water and soil samples show an increase of salinity over time in the region. Climate change induced sea-level rise will further intensify the problem of river and soil salinisation (World Bank 2016).

5.4.5 Marginalisation of Traditional Forest Users

The current management framework of the Sundarbans excludes the traditional forest resource users in the management process. Here exclusion means that the communities cannot apply their customary knowledge to resource management. Their exclusion from managing this forest led them to undermine the process of conservation because of inadequate representation of their interests. Moreover, the current management practice does not include alternative livelihood options for them.

5.5 Informal Institutions, Traditional Knowledge and Human Sociality: Towards Sustainable Conservation of Biodiversity Resources

The IPLCs sensibly believe that the forest provides their livelihoods, and it must be protected from all sorts of misuse and abuse for the present and future generations. They, therefore, follow some rules according to which they harvest the resources with utmost care and love for the nature (Fig. 5.10).

Fig. 5.10 Traditional rules and practices followed by IPLCs occupational groups at a glance. (Source: prepared by the authors)

5.5.1 Traditional Rules and Practices Followed by IPLCs

5.5.1.1 Rules Followed by the Mouals (Honey/Wax Collectors)

Honey is considered as an important non-wood forest product. The *Mouals* (honey/wax collectors), while collecting honey from the honeycombs, usually during the months of April, May and June, cut a specific section (about two thirds) of the honeycomb and leave the rest for reproduction. They also try to make sure that no young bees are killed while collecting honey and squeeze beehives by hand and never use metal tools. They revisit the colonies after a period of 1 month or more depending upon the size of the colony and flowering condition of nearby vegetation. When collecting the honey, the *Mouals* produce smoke using dry leaves but never put fire on beehive.

5.5.1.2 Rules Followed by Bawalis (Wood Collectors)

The *Bawalis* (wood collectors) follow several rules to ensure sustainable harvests of wood. They leave at least one stem in each clump of trees after cutting. Once the *Bawalis* have harvested wood from a compartment, in the following year they will not use this compartment for harvesting but will harvest on a cyclical basis so that there is an adequate re-growth of plants. They usually cut wood where there is abundance. They do not cut young and straight trees. The *Bawalis* believe that this tidal forest is a sacred place and the Creator washes the forest twice a day and maintains its sanctity and, therefore, try to maintain sustainable use of forest.

5.5.1.3   Traditional Practices of Golpata (*Nypa fruticans*) Harvesters

According to the rules followed by *Golpata* harvesters, exploitation in any area is not allowed more than once in a year and is not allowed during June to September specifically as it is the growing period of *Golpata (Nypa fruticans)*. They cut only the leaves that are approximately 9 ft long, and the leaves are cut in a way so that the central leaf and the leaf next to it in each clump are retained. They maintain the rule that the flowers and fruits shall in no way be disturbed when cutting leaves. They also maintain that young plants with only one utilisable leaf should not be cut.

5.5.1.4   Customary Rules Followed by Jele (Traditional Fishers)

The *Jele* (traditional fishers) knows that catching fish fry will ultimately deplete the number of fishes in the water bodies and thus they try to avoid doing so. They do not use '*jal*' net (very small-meshed net) usually. They use nets like *behundi jaal* (bag net) or *charpaataa* and *khaal-paataa jaal* (stake nets)—which are innovated and customised scientifically to benefit the Sundarban's unique waterscape. They use big-meshed net for rivers and small-meshed net for closed water bodies. They do not catch all species of fish and also avoid fishing in the spawning period.

5.5.2   Innovation and Diversification of Livelihood Patterns

In addition to the above-discussed traditional rules and practices which have been practiced through generations, the IPLCs in recent times have also diversified their livelihoods options by innovating different production methods and techniques as responses to the continuous deterioration of their livelihood opportunities due to man-made pressures (e.g. degradation of forest resources, loss of agricultural lands) and anthropogenic pressures such as climate change. These techniques are innovative as the IPLCs came up with these for enriching their adaptation capacity to the changed situation.

5.5.2.1   Innovative Techniques in Agriculture

The local small farmers have developed some innovative techniques in agriculture that are adaptive to local biophysical conditions while ensuring environmental sustainability. In the face of climate change and increased salinity in soil and water in that coastal region, the farmers grow their rice seedlings in raised land to reduce the risk of saline water contamination for ensuring maximum survival and then these seedlings are transplanted in the main agricultural land. For instance, they harvest rice plant at 8–12-in. high from the ground to respond to high salinity contents in soil and water (Fig. 5.11a). Practically this saline contaminated rice straw is decom-

Fig. 5.11 (**a**) Rice harvesting in raised lands and (**b**) cultivating vegetables on roof

posed within very short time if these are used as roofing materials. They, therefore, let those to be decomposed in the field which in turn add organic matter, mainly nitrogen, in soil and also reduce saline intensity, which is beneficial for the growth of their next crop. Moreover, those who are landless, grow vegetables on sheds or roofs, yard or back yard of their houses (Figure 5.11b).

5.5.2.2 Community-Based Mangrove Agro Aqua Silvi (CMAAS) Culture

The CMAAS culture refers to the practice of integrated cultivation of some mangrove faunal species—crabs, oyster or fishes (e.g. shrimps and bhetki [*Lates calcarifer*]) and floral species—golpata (*Nypa fruticans*), keora (*Sonneratia apetala*), goran (*Ceriops decandra*), etc. at the same time on any swampy land of brackish water. In addition, integrated cultivation of some mangrove floral species like *golpata* and a few faunal species like *tengra* (*Mystus tengara*), *baila* (*Awaous guamensis*), *tilapia* (*Tilapia nilotica*), etc. are practiced in a fresh water swampy land. The CMAAS culture is found to be profitable as is depicted in Table 5.3.

CMAAS culture is in fact an alternative practice to the commercial shrimp (CS) culture which has negligible or no adverse impact on the Sundarbans ecosystem. It has been pointed out already in the previous discussion that the commercial shrimp cultivation is leaving huge adverse impacts on the Sundarbans. Here, a comparative analysis of these two types of culture is provided in summary based on the findings of a research of *Unnayan Onneshan*.

The comparison in economic terms[3] can be depicted in Table 5.4. In terms of net present value (NPV) and net benefit (NB), CMAAS culture looks more profitable than commercial shrimp (CS) culture. But the scenario is quite different when considering benefit–cost ratio (BCR). The BCR scenario implies that the cost effectiveness of CS culture is comparatively higher. Shrimp cultivation is, therefore, no

---

[3] The cost–benefit analysis (CBA) approach was used to compare the economic returns in this case.

Table 5.3 Economic return of CMAAS culture (Source: prepared based on findings of the research by UO 2010)

CMAAS culture		
Economic return (Benefits > cost)	Mangrove cultivation (flora): Total income (per 'Bigha'/per year): BDT 56,250 Total cost (per 'Bigha'/per year): BDT 1800 Net benefit: BDT 54,450 Cost–benefit ratio: 1:32	Mangrove aqua farming (fauna): Total income (per 'Bigha'/per year): BDT 1,83,000 Total cost (per 'Bigha'/per year): BDT 14,750 Net benefit: BDT 173,250 Cost–benefit ratio: 1:12

Note: A *Bigha*, a unit of land measurement, is 1600 yd^2 (0.1338 hectare or 0.3306 acre) and often interpreted as being 1/3 acre (it is precisely 40/121 acre). In metric units, a bigha is hence 1333 m^2

Table 5.4 Value of cost–benefit analysis (CBA) measures of CMAAS and CS culture (Source: prepared based on findings of the research by *Unnayan Onneshan* 2010)

Measures of CBA	CMAAS culture(BDT/*bigha*/ year)	CS culture(BDT/*bigha*/ year)
Present value of costs (PVC)	16,550.00	8860.00
Present value of benefits (PVB)	217,500.00	177,272.72
Net present value (NPV)	202,454.54	169,218.18
Net benefit (NB)	200,950.00	168,412.72
Benefit–cost ratio (BCR)	13.00	20.00

doubt profitable. But beneficiaries are a selected group of people, and regrettably it has badly affected the livelihoods of landless and marginal farmers. Moreover, the ecological comparison (Table 5.5) proves that the CS culture is highly detrimental to the environment, whereas CMAAS culture has negligible or no harmful impact on the environment.

The ecological benefits resulting from the practice of CMAAS culture signify that the culture protects lands and soil from erosion, ensures better utilisation of fallow lands, protects environment from pollution, helps conserve biodiversity resources of the Sundarbans and most importantly provides alternative and sustainable livelihood options for the IPLCs.[4]

The CMAAS culture, as a whole, therefore, is a unique adaptation method to adapt to climate change in the coastal region. The local communities have invented this method, displaying a strong sense of ownership and a scope for scalability.

---

[4]The research, conducted by *Unnayan Onneshan,* focused only on a comparative analysis of CMAAS and CS culture based on economic and ecological indicators and has found it as a sustainable livelihood option for the IPLCs. More research can be conducted on a rigorous basis to assess its viability as an alternative income source for a wider context of coastal region for increased number of populations.

Table 5.5 Ecological Comparison between CMAAS and CS culture (Source: prepared based on findings of the research by *Unnayan Onneshan* 2010)

Criteria	CMAAS culture	CS culture
Salinity	No use of saline water; no salinity intrusion	Increases salinity in soil (in farmland and in adjacent lands)
Use of lands	Homestead adjacent fallow lands are used, and no conversion of forest lands into cultivation lands	Used ponds exhaust usefulness within 3–6 years of construction. So, destruction of mangroves occurs to make room for more ponds
Use of chemical fertiliser, pesticides, insecticides	No usage of chemical fertiliser or insecticides, natural feeding, and therefore no pollution	Chemical fertiliser, insecticides, etc. are used, causing pollution
Impact on agricultural productivity	Does not affect the agricultural productivity	Restricts crop production in agricultural land (by increasing salinity of lands) and conversion of agricultural lands to shrimp farming ponds reduces land availability
Impacts on the Sundarbans (in particular)	Eases and reduces the increasing anthropogenic pressures, making an alternative source of livelihoods for the local people who are dependent on the Sundarbans	Eradication of natural mangrove vegetation, and pollution of aquatic resources (negative)
Adaptation to climate change	An innovative adaptation method to climate change for the vulnerable	Increases the vulnerability to climate change

5.6   IPLCs, Resilience and Aichi Biodiversity Targets

As a Contracting Party to the Convention on Biological Diversity (CBD), Bangladesh is committed to implementing conservation and sustainable management of its biological diversity. The findings based upon empirical analysis, however, reveal that the most important biodiversity hotspot of this country, the Sundarbans, is under the threat of continuous degradation. In this process, the lives and livelihood conditions of the IPLCs are also being adversely impacted. Moreover, traditional knowledge-based livelihood strategies of the IPLCs are found to be effective in maintaining sustainable utilisation and conservation of this forest ecosystem. Yet, their knowledge has been neglected often under the formal institutional management system. Under the considerations of such major findings, this section *firstly* assesses the resilience capacity of the Sundarbans as a SEPLS based on some of the notable resilience indicators[5] considering two scenarios: (a) resilience capacity under current management process and (b) change in resilience capacity under the alternative conservation framework (developed in Sect. 5.3). A multiple evidence-based approach for the

---

[5] A set of indicators of resilience of SEPLS has been developed by UNU-IAS to provide a tool for communities to understand their resilience and encourage the practices that strengthen it (UNU-IAS 2015). In total 20 indicators are developed so far, but here some of the important indicators have been used to assess the case of the Sundarbans.

assessment of the resilience capacity has been followed. The findings of the assessment have, then, been summarised in Table 5.6 through triangulation of conceptual framework (developed through critical analysis of available secondary literature on natural resource management), primary data collected from the IPLCs through numerous consultations and authors' own interpretations on the former. In this regard, a significant amount of primary data has been collected through participatory approaches (Focus Group Discussions—FGD, unstructured interview, Participatory Rural Appraisal—PRA tools like social mapping, impact assessment by the respondents, etc.) particularly drawing on from knowledge, views and understandings of IPLCs who are the members of the three cooperatives that the *Unnayan Onneshan* had helped set up—*Harinagar Bonojibi Bohumukhi Unnayan Samity*, *Koyra Bonojibi Bohumukhi Unnayan Samity* and *Munda Adivasi Bonojibi Bohumukhi Unnayan Samity* in the adjacent regions of the Sundarbans (Fig. 5.12).

Secondly, the section also illustrates how the alternative measures as suggested by this study for ensuring sustainability of biodiversity of the Sundarbans can help achieve the Aichi Biodiversity Targets[6] envisioned by CBD.

A comparative analysis shows that human sociality-based alternative framework contributes significantly to the conservation of the Sundarbans biodiversity by making more resilient ecological system and society. This conservation practice directly impacts on 12 resilient indicators indicating a positive relationship (Table 5.6). It signifies that this framework is more ecologically responsive regarding the context of a SEPLS. For instance, under the current management approach, the ecosystem is hardly protected, and the regeneration capacity is hampered because of failure of checking anthropogenic pressures. On the contrary, the alternative framework tries to ensure the protection of the ecosystem at a higher level and revitalise the regeneration capacity at the fullest (indicator 1, 2, 3). This is possible as the alternative one puts high emphasis on the importance of the traditional knowledge system, whereas the current regime does not fully recognise the traditional knowledge (indicator 5, 6). In terms of the governance and equity indicators, the community-based governance is only envisioned in the policy paper, but in practice such governance system is undermined by agencies of the government. The alternative suggestions, on the other hand—the participation of the community in resource management—build a social capital that contributes to the cooperation, social equity and efficient governance (indicator 7, 8, 9, 10). Both the management frameworks (current and alternative) recognise that the livelihoods of the local people are based on biodiversity resources of the Sundarbans (indicator 12). The alternative framework, however, emphasises that this biodiversity-based livelihood pattern should be maintained in a sustainable way that conserves the biodiversity resources (indicator 4) as well as provides alternative livelihoods under the changed circumstances by diversifying their income sources (indicator 11).

---

[6]A set of 20 global targets under the Strategic Plan for Biodiversity 2011–2020 (CBD 2013; CBD Secretariat 2014).

Table 5.6 Comparative analysis of resilience capacity of the Sundarbans under two different scenarios (Source: prepared by the authors)

Resilience indicators	Scenario under current practice of management					Scenario under the alternative conservation framework				
	Very high	High	Medium	Low	Very low	Very high	High	Medium	Low	Very low
Landscape and seascape diversity and ecosystem protection										
1. Ecosystem protection				√		√				
2. Ecological interaction considered				√		√				
3. Recovery and regeneration					√	√				
Biodiversity										
4. Sustainable management of biodiversity resources					√	√				
Knowledge and innovation										
5. Traditional knowledge related to biodiversity			√			√				
6. Documentation of biodiversity-associated knowledge					√	√				
Governance and social equity										
7. Rights of the community in resource management				√		√				
8. Community-based governance				√		√				
9. Social capital as cooperation and coordination in resource management					√	√				
10. Social equity					√		√			
Livelihood and well-being										
11. Income diversity				√			√			
12. Biodiversity-based livelihoods	√					√				

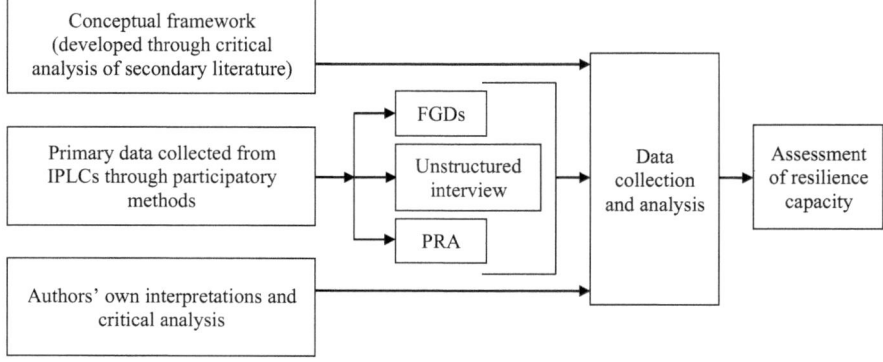

Fig. 5.12 Methods followed for the collection and analysis of data for the assessment of the resilience capacity of the Sundarbans. (Source: prepared by the authors)

Table 5.7 Achievement of Aichi Biodiversity Targets under the alternative conservation framework

Targets	Relevant indicators/issues	Contribution of this case study
Target 10: Pressures on vulnerable ecosystems reduced	• Trends in extent, of vulnerable ecosystems (here mangrove) • Anthropogenic pressures • Climate change	• Multiple anthropogenic pressures identified on a mangrove ecosystem • Presenting and promoting the TK-based climate adaptation methods and sustainable agricultural methods
Target 15: Ecosystem restored and resilience enhanced	• Ecosystem resilience • Restoration	• Traditional rules and methods followed by IPLCs promote the restoration process and enhances resilience capacity • Climate change adaptation methods like CMAAS innovated by the IPLCs enhances resilience capacity
Target 18: Traditional knowledge respected	• Traditional knowledge, innovations and practices • Customary use of biological resources	• Promotes TK knowledge system practised by the IPLCs • Urges to recognise the traditional practices in the resource management framework • Emphasises on the participation of IPLCs in the resource management

The alternative conservation framework, accordingly, helps achieve some of the important targets under 'Aichi Biodiversity Targets' as is illustrated in Table 5.7. Firstly, it helps to contribute to the Target no. 10 by reducing pressures on vulnerable (here, mangrove) ecosystem. Secondly, it promotes restoration and enhanced resilience of that ecosystem and thus helps achieve Target no. 15. Finally, and most importantly, it contributes to achieve Target no. 18 by respecting the TK system practised by the local and indigenous communities (Table 5.7).

5.7 Conclusions

There is a significant number of anthropogenic pressures that cause the degradation of biodiversity resources of the Sundarbans. These anthropogenic pressures have mainly intensified with the advent of neo-liberalism as the sole strategy of accumulation of wealth, with profits being considered more important through commercialisation of forest products, neglecting intrinsic ecological value of biological resources. These commercial enterprises, formal and informal, are found to be highly organised in their extractions of resources, and most often being politically patronised and administratively supported. The chapter, thereafter, has scrutinised the livelihood strategies of the IPLCs, the resource-dependent communities of the Sundarbans, and the results show that their livelihood strategies (both traditional practices and innovative tools) are largely effective and beneficial for the protection and maintenance of natural mangrove ecosystem. The assessment of the Sundarbans on the basis of the resilience indicators of SEPLS also shows that the current resilience capacity can be improved by mainstreaming the traditional knowledge base and participation of the indigenous people into the resource management framework.

The lessons from this study can be applied with necessary modifications to improve policy decisions and management interventions of such type of SEPLS in different countries of the world. There is no denying of the necessity to revise laws, regulations, and policies relating to the use of resources and to secure the rights of the IPLCs.

Bibliography

Adhikari KP, Goldey P (2010) Social capital and its "downside": the impact on Sustainability of induced community-based organizations in Nepal. World Dev 38(2):184–194

Aziz A, Paul AR (2015) Bangladesh Sundarbans: present status of the environment and biota. Diversity 7(3):242–269

Bangladesh Bureau of Statistics (BBS) (2001) Population and housing census 2001. Government of the People's Republic of Bangladesh, Dhaka

Bangladesh Bureau of Statistics (BBS) (2011) Population and housing census 2011. Government of the People's Republic of Bangladesh, Dhaka

Bangladesh Forest Department (BFD) (2015) First phase Tiger status report of Bangladesh Sundarban. Wildlife Institute of India and Bangladesh Forest Department, Ministry of Environment and Forests, Government of the People's Republic of Bangladesh, Dhaka

Barbier EB (2007) Valuing ecosystem services as productive inputs. Econ Policy 22:177–229

Baten MA, Kumar U (2010) Responses to the changes in the Sundarbans. Paper presented at international conference on biological and cultural diversity, UNCBD, Montreal, Canada, 3–5 May

Battersby S (2017) Can humankind escape the tragedy of the commons? Proc Natl Acad Sci 114(1):7–10

Behera MD, Haider MS (2012) Situation analysis on biodiversity conservation–ecosystem for life—a Bangladesh India Initiative. IUCN, Dhaka

Bergamini N, Blasiak R, Eyzaguirre P, Ichikawa K, Mijatovic D, Nakao F, Subramanian SM (2013) Indicators of resilience in socio-ecological production landscapes (SEPLs), UNU-IAS Policy Report

Bräuer I (2003) Money as an indicator: to make use of economic evaluation for biodiversity conservation. Agric Ecosyst Environ 98:483–491

CBD (2013) Quick guides to the Aichi Biodiversity Targets - version 2. https://www.cbd.int/doc/strategic-plan/targets/compilation-quick-guide-en.pdf

CBD Secretariat (2014) Aichi Biodiversity Targets. https://www.cbd.int/sp/targets/

CEGIS (2016) Joint landscape narrative by India and Bangladesh. Center for Environmental and Geographic Information Services (CEGIS), Unpublished Report

Costanza R, d'Arge R, de Groot R, Farberk S, Grasso M, Hannon B, Limburg K, Naeem S, O'Neill RV, Paruelo J, Raskin RG, Sutton P, van den Belt M (1997) The value of the world's ecosystem services and natural capital. Nature 387:253–260

DoE (2015) Fifth national report to the convention on biological diversity: biodiversity national assessment and programme of action 2020. Ministry of Environment and Forest, Government of the People's Republic of Bangladesh

FAO (2011) Bangladesh forestry outlook study. Asia-Pacific forestry sector outlook II. Working paper series, paper no. APFSOS II/WP/2011/33, p 101

Giri C, Pengra B, Zhu Z, Singh A, Tieszen LL (2007) Monitoring mangrove forest dynamics of the Sundarbans in Bangladesh and India using multitemporal satellite data from 1973 to 2000. Estuar Coast Shelf Sci 73:91–100

Giri C, Long J, Abbas S, Murali RM, Qamer FM, Pengra B, Thau D (2014) Distribution and dynamics of mangrove forests of South Asia. J Environ Manag 148:1–11

Gomez-Baggetthun E, Ruiz-Perez M (2011) Economic valuation and the commodification of ecosystem services. Prog Phys Geogr 35:613–628

Gopal B, Chauhan M (2006) Biodiversity and its conservation in the Sundarbans mangrove ecosystem. Aquat Sci 68:338–354

Greenberg JB, Park TK (1994) Political ecology. J Polit Ecol 1:1–12

Gu H, Subramanian SM (2012) Socio-ecological production landscapes: relevance to the green economy agenda. UNU-IAS Policy Report. UNU-IAS, Japan

Hahn T, McDermott C, Ituarte-Lima C, Schultz M, Green T, Tuvendal M (2015) Purposes and degrees of commodification: economic instruments for biodiversity and ecosystem services need not rely on markets or monetary valuation. Ecosyst Serv 16:74–82

Higgins P, Short D, South N (2013) Protecting the planet: a proposal for a law of ecocide. Crime Law Soc Chang 59:251–266

Hussain MZ (2014) Bangladesh Sundarban delta vision 2050: a first step in its formulation – document 2: a compilation of background information, IUCN, International Union for Conservation of Nature, Bangladesh Country Office. Dhaka, Bangladesh, viii+192 pp

Ichikawa K (2013) Understanding socio-ecological production landscapes in the context of Cambodia. Int J Environ Rural Dev 4(1):57–62

Iftekhar MS, Islam MR (2004) Degeneration of Bangladesh's Sundarbans mangroves: a management issue. Int For Rev 6(2):123–135

Islam MSN, Gnauck A (2009) Threats to the Sundarbans mangrove wetland ecosystems from Transboundary water allocation in the Ganges Basin: a preliminary problem analysis. Int J Ecol Econ Stat 13(9):64–78

Islam MT (2014) Vegetation changes of Shundarbans based on Landsat imagery analysis between 1975 and 2006. Landsc Environ 8(1):1–9

Ituarte-Lima C, Schultz M, Hahn MC, Cornell S (2014) Biodiversity financing and safeguards: lessons learned and proposed guidelines. Information Document UNEP/CBD/COP/12/INF/27 for the 12th conference of the parties of the convention on biological diversity in Pyeongchang Korea. SwedBio/Stockholm Resilience Centre at Stockholm University, Stockholm

IUCN Bangladesh (2014) Bangladesh Sundarban Delta Vision 2050: A first step in its formulation - Document 1: The Vision, Dhaka, Bangladesh: IUCN, vi+23 pp. https://portals.iucn.org/library/sites/library/files/documents/2014-065-doc.1.pdf

Kabir DMH, Hossain J (2008) Resuscitating the Sundarbans: customary use of biodiversity and traditional cultural practices in Bangladesh. Unnyan Onneshan, BELA, Forest Peoples Programme and Nijera Kori, Dhaka

Kosoy N, Corbera E (2010) Payments for ecosystem services as commodity fetishism. Ecol Econ 69:1228–1236

Kumar P (2005) Market for ecosystem services. International Institute for Sustainable Development (IISD), Winnipeg

Liu JJ, Harris L, Zhao HJ, Qian F (2012) Integrating community development with the management of grasslands at Ke'erqin Nature Reserve, Inner Mogolia, China. In IPSI Secretariat, IPSI case study booklet. United Nations University: Institute for the Advanced Study of Sustainability, Yokohama

McAfee K, Shapiro EN (2010) Payments for ecosystem services in Mexico: nature, neoliberalism, social movements, and the state. Ann Assoc Am Geogr 100:579–599

Muradian R, Arsel M, Pellegrini L, Adaman F, Aguilar B, Agarwal B, Corbera E, Ezzine de Blas D, Farley J, Froger G (2013) Payments for ecosystem services and the fatal attraction of win-win solutions. Conserv Lett 6:274–279

Neuteleers S, Engelen B (2015) Talking money: how market based valuation can undermine environmental protection. Ecol Econ 117:253–260

Ostrom E (2000) Private and common property rights. Center for the Study of Institutions, Population and Environmental Change, Indiana University, Bloomington

Pearce D (2001) Valuing biological diversity: issues and overview. In: Valuation of biodiversity benefits: selected studies. OECD, Paris

Perrings C, Folke C, Mäler K (1992) The ecology and economics of biodiversity loss: the research agenda. Ambio 21(3):201–211

Rahman MM, Rahman MM, Islam KS (2010) The causes of deterioration of Sundarbans mangrove forest ecosystem of Bangladesh: conservation and sustainable management issues. Aquac Aquar Conserv Legis Int J Bioflux Soc 3(2):77–90

Rahman MR, Asaduzzaman M (2010) Ecology of Sundarbans, Bangladesh. J Sci Foundation 8(1&2):35–47

Robbins P (2012) Political ecology: a critical introduction, 2nd edn. Blackwell, West Sussex

Roy AKD, Alam K (2012) Participatory forest management for the sustainable management of the Sundarbans mangrove forest. Am J Environ Sci 8(5):549–555

Sadmo A (2015) The early history of environmental economics. Rev Environ Econ Policy 9(1):43–63

Sarker SK, Reeve R, Thompson J, Paul NK, Matthiopoulos J (2016) Are we failing to protect threatened mangroves in the Sundarbans world heritage ecosystem? Scientific Reports

Schlager E, Ostrom E (1992) Property rights regimes and natural resources: a conceptual analysis. Land Econ 68(3):249–262

The Guardian (2015) Only 100 tigers left in Bangladesh's famed Sundarbans forest, 27 July 2015. https://www.theguardian.com/environment/2015/jul/27/only-100-tigers-left-in-bangladeshs-famed-sundarbans-forest

The Daily Star (2018) Licence to harm Sundarbans, 6 April 2018. https://www.thedailystar.net/frontpage/licence-harm-sundarbans-1558918

Titumir RAM, Afrin T (2017) Complementarities of human-nature well-beings: a case illustrated through traditional forest resource users of Sundarbans in Bangladesh. Satoyama Initiative Thematic Review (SITR), vol 3. United Nations University Institute for the Advanced Study of Sustainability (UNU-IAS), Tokyo

Titumir RAM (2011) Property rights, CSU and development. Paper presented at the Expert Committee meeting on CBD article 8(j), CBD Secretariat, Montreal

Titumir RAM (2015) IPLCs contribution to Aichi Biodiversity Target 10: a case illustrated through TRUs of Sundarbans in Bangladesh. Paper presented at side event on indigenous peoples and local communities' contributions to the implementation of the strategic plan for biodiversity 2011–2020, Montreal, 6 Nov 2015

Titumir RAM, Afrin T, Islam MS (In progress) Well-being of nature: biodiversity, water resource and climate change in Bangladesh Context. Unpublished book, UnnyanOnneshan, Dhaka

Turnhout E, Waterton C, Neves K, Buizer M (2013) Rethinking biodiversity: from goods and services to "living with". Conserv Lett 6:154–161

Uddin MS, Shah MAR, Khanom S, Nesha MK (2013) Climate change impacts on the Sundarbans mangrove ecosystem services and dependent livelihoods in Bangladesh. Asian J Conserv Biol 2(2):152–156

Unnayan Onneshan (UO) (2010) Community based mangrove agro aqua silvi (CMAAS) culture: promoting as a community adaptation tool and an alternative to commercial shrimp culture. Research report. Unnayan Onneshan, Dhaka

UNU-IAS (2015) Indicators of resilience in socio-ecological production landscapes and seascapes, 2nd edn. United Nations University Institute for the Advanced Study of Sustainability (UNU-IAS), Tokyo

Vatn A (2009) An institutional analysis of methods for environmental appraisal. Ecol Econ 68:2207–2215

Vatn A (2010) An institutional analysis of payments for environmental services. Ecol Econ 69:1245–1252

World Bank (2016) Impact of climate change and aquatic salinization on mangrove species and poor communities in the Bangladesh Sundarbans. World Bank Policy Research Paper 7736

Chapter 6
Lessons Learned from Application of the "Indicators of Resilience in Socio-ecological Production Landscapes and Seascapes (SEPLS)" Under the Satoyama Initiative

William Dunbar, Suneetha M Subramanian, Ikuko Matsumoto, Yoji Natori, Devon Dublin, Nadia Bergamini, Dunja Mijatovic, Alejandro González Álvarez, Evonne Yiu, Kaoru Ichikawa, Yasuyuki Morimoto, Michael Halewood, Patrick Maundu, Diana Salvemini, Tamara Tschenscher, and Gregory Mock

W. Dunbar (✉) · E. Yiu
United Nations University Institute for the Advanced Study of Sustainability, Tokyo, Japan
e-mail: dunbar@unu.edu

S. M. Subramanian
United Nations University International Institute for Global Health (UNU-IIGH), Cheras, Kuala Lumpur, Malaysia

I. Matsumoto
Natural Resources and Ecosystem Services Area, Institute for Global Environmental Strategies (IGES), Hayama, Kanagawa, Japan

Y. Natori · D. Dublin
Conservation International Japan, Tokyo, Japan

N. Bergamini · D. Mijatovic · Y. Morimoto · M. Halewood
Bioversity International, Rome, Italy

A. G. Álvarez
Instituto de Investigaciones Fundamentales en Agricultura Tropical, Havana, Cuba

K. Ichikawa
Institute of Policy Research, Kumamoto, Japan

P. Maundu
Kenya Resource Centre for Indigenous Knowledge, The National Museums of Kenya, Nairobi, Kenya

D. Salvemini · T. Tschenscher
United Nations Development Programme, New York, NY, USA

G. Mock
Independent Contractor, Seattle, WA, USA

© The Author(s) 2020
O. Saito et al. (eds.), *Managing Socio-ecological Production Landscapes and Seascapes for Sustainable Communities in Asia*, Science for Sustainable Societies, https://doi.org/10.1007/978-981-15-1133-2_6

Abstract Socio-ecological resilience is vital for the long-term sustainability of communities in production landscapes and seascapes, but community members often find it difficult to understand and assess their own resilience in the face of changes that affect them over time due to economic and natural drivers, demographic changes, and market forces among others, due to the complexity of the concept of resilience and the many factors influencing the landscape or seascape. This chapter provides an overview of a project and its resilience assessment process using an indicator-based approach, which has been implemented under the International Partnership for the Satoyama Initiative (IPSI). In this project, a set of 20 indicators were identified to capture different aspects of resilience in SEPLS, and examples are included from various contexts around the world, with the purpose of identifying lessons learned and good practices for resilience assessment. These indicators have now been used by communities in many countries, often with the guidance of project implementers, with the goal of assessing, considering, and monitoring their landscape or seascape's circumstances, identifying important issues, and ultimately improving their resilience. While this particular approach is limited in that it cannot be used for comparison of different landscapes and seascapes, as it relies on community members' individual perceptions, it is found useful to understand multiple aspects of resilience and changes over time within a landscape or seascape.

Keywords Satoyama Initiative · Socio-ecological production landscapes and seascapes (SEPLS) · Indicators · Resilience · Assessment

6.1 Introduction and Background

The "Indicators of Resilience in Socio-ecological Production Landscapes and Seascapes (SEPLS)" are a set of 20 indicators for communities to assess the socio-ecological resilience of the production landscapes and seascapes on which they rely for their livelihoods and well-being. While socio-ecological resilience is a complex concept, for the purposes of this project, it is considered that resilience refers to "the capacity of a system to deal with change and continue to develop; withstanding shocks and disturbances and using such events to catalyze renewal and innovation" (Stockholm Resilience Center 2014). The set of indicators has been piloted, field-tested and applied over nearly 10 years through a number of programs, some of which are introduced in this chapter, with the result that communities in the landscapes and seascapes covered by these programs have better understood their own resilience and developed strategies for improvement. The indicators also have an added benefit as a capacity-building tool, as the process of using them for resilience assessment helps local community members to understand how they can be actively involved in resilience improvement through actions on the ground and to learn about concepts that are important for planning activities and project design, but are often unfamiliar to farmers, fishers, and other ground-level practitioners.

Resilience assessment stimulates active dialogue among actors with diverse attributes and backgrounds.

The background of the indicators dates back to the beginnings of the Satoyama Initiative, a global initiative to realize its vision of "societies in harmony with nature" through the revitalization and sustainable management of SEPLS. The Satoyama Initiative was established based heavily on research results from the "Japan *Satoyama Satoumi* Assessment" (JSSA), a multi-year assessment carried out in Japan of *satoyama* and *satoumi,* which are Japanese landscapes and seascapes dominated by human production activities, i.e., Japanese SEPLS. Among the JSSA's findings were that: (1) these landscapes and seascapes are composed of an interlinked mosaic of ecosystem types that are managed to provide for human well-being; (2) they have undergone significant changes in recent years that have caused a drop in their resilience; (3) this trend has important consequences for human well-being and biodiversity; and (4) integrated approaches to address this trend have the potential to reduce biodiversity loss and maintain sustainable flows of ecosystem services (UNU-IAS 2010).

For these integrated approaches to be developed and implemented, a need was identified to first assess resilience in order to be able to maintain and strengthen it. Resilience in production landscapes and seascapes is a function of their dynamic and evolving ecological, social, cultural and economic systems, not of any static set of natural resource uses or species, making it impossible to measure precisely with any simple yardstick. With this complexity in mind, an initial set of indicators was developed jointly by Bioversity International and the United Nations University Institute for the Advanced Study of Sustainability (UNU-IAS). Much of the background and reasoning behind this process was compiled in the policy brief "Indicators of Resilience in Socio-ecological Production Landscapes (SEPLs)" produced by UNU-IAS in 2013 (Bergamini et al. 2013). Findings of the report showed that the indicators approach can help identify gaps in knowledge and areas of intervention to improve resilience in target communities. This report also identified several principles that informed the selection of the indicators, including that they should be easy to understand by local land users; that they should reflect the views of various stakeholders; and that people's perceptions and needs change over time. In addition to the findings of the JSSA and other research, the indicators were based on case studies collected under the Satoyama Initiative that demonstrated communities' abilities to build their resilience.

As cited above, the JSSA found that the interlinked nature of SEPLS—meaning interlinkages between people and nature, between different ecosystem processes, between ecosystem services and human well-being, and others—gives resilience to their socio-ecological production systems. This is what makes resilience difficult to measure, and also what previous research found pointed to the need for an indicators approach that considers the social and cultural dimensions of ecosystem functioning including temporal changes (van Oudenhoven et al. 2011). Related research has found that community-level resilience encompasses a diversity of ecological, socioeconomic, and other variables, suggesting that an integrated model could be used for assessment of resilience based on a matrix of these variables (Antwi et al. 2014). The indicators were therefore designed to allow local communities to monitor social dimensions in addition to ecological factors, and also to help implement and evaluate conservation approaches, as informed by the Satoyama Initiative case studies (Bergamini et al. 2013). Because this indicator approach is based on community members' perceptions, it is limited in that it cannot be used for comparison of different SEPLS, rather collecting subjective information for the use of the community itself.

After the initial set of indicators was field-tested and applied in projects in over 20 countries, the indicators were further refined and updated, and a "Toolkit" publication was published to facilitate their use in 2014 (UNU-IAS et al. 2014). The toolkit provides a revised set of the 20 indicators, practical guidance on how to use them for resilience assessment, and examples of their use from the field. The current set of indicators and methodology are now being used by projects working to improve resilience in communities around the world. This chapter provides an overview of the indicators, some examples of how they are used in projects, and findings from these processes and projects.

6.2 The Indicators and Resilience Assessment

As mentioned above, the 20 indicators have been selected to help communities assess the resilience of the socio-ecological systems in the landscapes and seascapes on which they rely for their well-being. The indicators are grouped into five areas, outlining practices and institutions that contribute to resilience in SEPLS and account for the specific social and ecological functions and components that make up the SEPLS system as follows:

Landscape or seascape diversity and ecosystem protection

1. Landscape/seascape diversity
2. Ecosystem protection
3. Ecological interactions between different components of the landscape/seascape
4. Recovery and regeneration of the landscape/seascape

Biodiversity (including agricultural biodiversity)

5. Diversity of local food system
6. Maintenance and use of local crop varieties and animal breeds
7. Sustainable management of common resources

Knowledge and innovation

8. Innovation in agriculture and conservation practices
9. Traditional knowledge related to biodiversity
10. Documentation of biodiversity-associated knowledge
11. Women's knowledge

Governance and social equity

12. Rights in relation to land/water and other natural resource management
13. Community-based landscape/seascape governance
14. Social capital in the form of cooperation across the landscape/seascape
15. Social equity (including gender equity)

Livelihoods and well-being

16. Socioeconomic infrastructure
17. Human health and environmental conditions
18. Income diversity
19. Biodiversity-based livelihoods
20. Socio-ecological mobility

Each of the 20 indicators listed above is provided in the toolkit publication (UNU-IAS et al. 2014) with a description, examples where appropriate, a question to be asked in assessing the indicator, explanations of high and low scores, and additional discussion questions where appropriate. The indicators are intended to be scored by individual participants first, then collectively among all participants, on a scale from 1 to 5, with 1 meaning the situation is least likely to be conducive to resilience, and 5 meaning the most favorable situation. For example, for the first indicator, "landscape/seascape diversity," a score of 1 would mean an extremely low level of diversity of natural ecosystems and land uses in the landscape or seascape, while a score of 5 would indicate high diversity, considered likely to contribute to resilience. In many cases, the temporal trend of the indicator may also be assessed, indicating whether the situation is perceived to be improving, deteriorating, or unchanging.

While anyone can use these indicators in whatever way and for whatever purpose they like, the process presented in the toolkit publication is a community-based resilience assessment workshop, which allows for an interactive and participatory process for community members to understand and discuss resilience. In these workshops, a representative group of landscape or seascape residents along with any other relevant stakeholders, with as broad as possible representation in order to ensure equity and diversity of voices, is invited to take part. The procedure of the workshop may vary depending on the purpose and intended outcomes of using the indicators. Generally, the agenda should include: an introduction to key concepts; explanation of the purpose of the workshop; exercises such as community mapping and/or creating historical timelines; scoring of the indicators themselves; and discussion of the results of the scoring. This process not only collects community members' opinions, but gives them a chance to consider the shape and conditions of their own landscapes and seascapes, and understand concepts and topics such as "biodiversity" and "resilience." These concepts may be more commonly used in academic or policymaking circles than among community members on the ground, but are important for all participants to understand in order to have an assessment that is accurate and based on common understanding. The dialogue that takes place during the assessment can reveal rich information on how stakeholders view their landscapes or seascapes similarly or differently, and stimulates discussion among actors who may not have regular interactions (e.g., elders and youth, different ethnicities, socioeconomic statuses).

Information and opinions on landscape and seascape resilience collected through this process have been used for a variety of purposes. Some examples are given in the next section of this chapter and include: to identify a baseline for producing a sustainable development strategy at the landscape/seascape level and design projects to implement the strategy, as seen in the cases in Sects. 6.3.1 and 6.3.2; for monitoring and evaluating the effectiveness of ongoing resilience-strengthening programs, as in Sects. 6.3.3 and 6.3.4; and as data for academic research projects, as in Sect. 6.3.5.

6.3 Experiences Using the Indicators of Resilience

6.3.1 Use of the Indicators to Facilitate Participatory Governance and Decision-Making: The COMDEKS Program

The "Community Development and Knowledge Management for the Satoyama Initiative" (COMDEKS) program has, since 2011, piloted a community-based model of landscape management in 20 landscapes and seascapes around the world, with the core objective to restore resilience in the face of a changing climate and socioeconomic challenges, protect biodiversity, and sustain SEPLS. The indicators are one of the principal tools employed by COMDEKS to gather information on current conditions and trends in different dimensions of resilience, link them to management practices past and present, and deepen community understanding of what these observations mean in relation to resilience. Repeated use of the indicators allows for adaptive management, where assessment results are used to continuously update activities in line with community needs. COMDEKS is implemented by the United Nations Development Programme (UNDP) in partnership with the Ministry of the Environment of Japan, the Secretariat of the Convention on Biological Diversity, and UNU-IAS, and is funded by the Japan Biodiversity Fund. The Global Environment Facility Small Grants Programme (GEF SGP) functions as its delivery mechanism and provides co-financing as well as technical and human resources to oversee its implementation.

The COMDEKS methodology relies on community consultation to drive a process of participatory landscape planning, and the indicators are central to the community consultation process. As part of this process, community members and other stakeholders come together to conduct a baseline assessment using the indicators, which is then used to define a "landscape strategy." Based on community perspectives and priorities delineated in the landscape strategy, projects in the community are identified and provided funding to implement the strategy. The indicators are integral to discussion, analysis, and negotiation in the processes of generating baseline information, reaching consensus on major challenges to local resilience, and developing a plan of action to address these challenges. Because of their effectiveness in promoting group discussion and interaction, they are also critical in generating the social capital necessary to undertake community-driven projects.

Under COMDEKS, target communities come together to discuss and score each indicator during a workshop organized as part of the baseline assessment of the landscape or seascape. This process is just as important for its educational purposes as for its role in generating data. Experience has shown that discussing and scoring the indicators acts as an effective introduction to the principles of landscape and seascape resilience. First, group discussion before scoring provides an opportunity to talk about resilience with local examples. The scoring exercise itself grounds this more general discussion in local experience, acting as a means to consider landscape conditions and trends and what they mean for resilience. Gaining an appreciation for the concept of resilience and how it manifests locally is one of the most important factors for the community in the early stages of the COMDEKS process.

The scores given for each indicator by stakeholders during the baseline assessment workshop provide essential input for the community to develop its landscape strategy. This is the most critical part of the planning process, where the community comes up with a vision of what a more resilient landscape would look like and determines what actions would be required to realize this vision. Although the scores are not quantitative measures of resilience, they do help identify potential problems that the strategy can address through COMDEKS projects.

An ex-post baseline assessment carried out at the completion of COMDEKS projects also uses the indicators to identify changes in resilience. A workshop similar to that carried out for the baseline assessment is held, at which the indicators are again scored by the community, and these scores are compared with the earlier ones. Although comparing the scores from the baseline assessment with those from the ex-post assessment cannot be used as a quantitative measure of landscape resilience change, it can be used to highlight local perceptions of changes due to the completed projects, and other factors affecting landscape resilience, and to indicate progress toward the goals identified in the landscape strategy and recommend adaptive measures. Thus, the indicators are an integral feature of COMDEKS implementation from beginning to end. They are also key to the adaptive management cycle that COMDEKS is based on, in which project results are used as a source of learning and innovation for future community efforts. The indicator scores, in addition to other progress metrics, are essential elements in monitoring and evaluation (M&E) processes.

The COMDEKS program has shown that besides facilitating a common understanding and vision for participatory landscape and seascape management among local stakeholders, the indicators can play an important part in giving community-level interventions legitimacy in the eyes of policy makers. By involving policy makers in the process and helping them to understand the elements that benefit resilience, and further by demonstrating that these elements can be considered and evaluated systematically, experience using the indicators has in some cases made it easier for policy makers to consent management actions designed to rebuild and sustain resilience.

Similarly, the indicators contribute greatly to resilience-focused SEPLS governance. They offer a method for assessing landscape changes as perceived by local landscape users, and evaluating landscape interventions as part of an adaptive management process. As such, they represent a potentially powerful tool for governance and sustainability planning. Community-based management actions based on their use have already proven effective in protecting local biodiversity while enhancing rural livelihoods and revitalizing local cultures in landscapes and seascapes covered under the COMDEKS program (UNDP 2018).

6.3.2   Using the Indicators for Community Benefits Under the "GEF-Satoyama Project"

The project "Mainstreaming Biodiversity Conservation and Sustainable Management in Priority Socio-ecological Production Landscapes and Seascapes" (or the "GEF-Satoyama Project") was developed with the aim of achieving societies in harmony with nature with a sustainable primary production sector based on traditional and modern wisdom, to make significant contributions to global targets for conservation of biological diversity. The project consists of three mutually interacting components: on-the-ground demonstration, with investments in ten subgrant projects in ten countries from the Indo-Burma, Tropical Andes, and Madagascar and Indian Ocean Islands Biodiversity Hotspots; knowledge generation through case studies and mapping; and capacity-building and awareness raising. The GEF-Satoyama Project is funded by the Global Environment Facility (GEF), implemented by Conservation International and executed by Conservation International Japan in cooperation with UNU-IAS and the Institute for Global Environmental Strategies (IGES).

The indicators were applied for baseline and progress monitoring under the GEF-Satoyama Project. All ten proponents of the site-based projects[1] conducted assessments using the indicators of resilience at the beginning and end of the implementation of their projects' interventions to document the status of landscape or seascape resilience. The assessment is designed as a participatory process that engages a variety of stakeholders including community members, civil society organizations, government agencies, and others. The use of the indicators enables the

---

[1] The subgrant proponents are: Asociación Amazónicos por la Amazonia (AMPA), Peru; Dahari, Comoros; Environmental Protection and Conservation Organization (EPCO), Mauritius; Fauna and Flora International (FFI), Myanmar; Fundación para la Investigación y Desarrollo Social (FIDES), Ecuador; Green Islands Foundation (GIF), Seychelles; Inter Mountain Peoples' Education and Culture in Thailand (IMPECT), Thailand; The Energy and Resources Institute (TERI), India; Universidad Industrial de Santander (UIS), Colombia; and Wildlife Conservation Society (WCS), Madagascar.

identification of priority actions for local innovation and implementation of adaptive management through community-led activities.

Toward these ends, training in the use of the indicators was provided to the project proponents, their partners, and other stakeholders, to enable them to contribute to building environmental and social resilience on the ground. Some of the key benefits gained from the application of the indicators in the GEF-Satoyama Project are discussed below.

- Recognizing the value of nature

- According to the results of the use of the indicators under the GEF-Satoyama Project, resilience assessment helped community members to re-evaluate nature in their communities through discussion of the diversity of flora, fauna, and food sources; the exercise of mapping their landscapes and seascapes; scoring of the indicators; and sharing and understanding their reasons for scoring them as they did. The discussion also helped local communities to revisit their history and socio-ecological and political conditions related to how they have engaged with their landscapes and seascapes, share their perceptions related to nature and their management, and recognize the evolution of their SEPLS. In turn, this encouraged more active participation of local communities in the development of participatory management plans, conservation practices, natural resource management, and enforcement. In the project in Zunheboto District of Nagaland in northeast India, conducted by The Energy and Resources Institute (TERI), the villagers of Sukhai, Ghukhuyi, and Kivikhu had become concerned that they were consuming too much wildlife, which led them to establish a community conserved area (CCA) in their territory. CCAs are areas voluntarily put into protection by communities for the purpose of conservation through consensus under self-imposed management rules. Although the ban on hunting has an economic impact on all community members due to loss of income from wildlife sales, after carrying out the resilience assessment they agreed that they could benefit more from conservation through activities such as ecotourism involving bird and butterfly watching. Activities also include making use of available natural resources by revitalizing their traditional weaving of handicrafts or shawls, a skill that only a few women currently possess. During the resilience assessment, members of the communities recognized that they still have extraordinarily rich biodiversity in their ecosystems. However, many of them found that unique ecosystems and the associated natural resources they were benefiting from were declining in size and quality, and understood the need to take more rigorous management measures to keep receiving these benefits. Thus, the resilience assessment helped them to identify challenges in their SEPLS, and encouraged active engagement of local communities to develop and implement natural resource management plans.

- Revitalizing indigenous and local knowledge

- The resilience assessments also evaluated local knowledge systems, including indigenous, local, and women's knowledge. This exercise helped local communities to recognize and take stock of their indigenous and local knowledge and how they utilize, maintain, and transfer the knowledge to the next generation. They also considered how they can innovatively use new knowledge integrated with traditional knowledge within their SEPLS. For instance, in a rainforest community in Makira Natural Park in northeastern Madagascar under the project conducted by Wildlife Conservation Society-Madagascar, people found that they are active in innovation in agriculture and conservation practices such as changes in intensive and improved rice cultivation systems using a "system of rice intensification" and "improved rice system"; establishment of permaculture and agroforestry; and reactivation of clove and cacao plantations to help against erosion. At the same time, they found that transmission of local knowledge still exists verbally, with elders having the impression that younger generations do not show much interest in learning about medicinal plants but rather prefer modern medicines. As a result of the resilience assessment workshop, local communities agreed to collect documents that provide knowledge about biodiversity, and to build a database to be used by school programs and for distribution of information.

- Similarly, most of the communities under the GEF-Satoyama Project were reminded of their indigenous, local, and women's knowledge through the assessment workshop and recognized that this knowledge was not being appropriately transferred to the next generation. Thus, many of them came up with some means for local documentation or improvement of communication among community members.

- Strengthening local governance and social equity

- The micro-watershed of the Las Cruces stream in the central Santander District in northwest Colombia is an area that was "opened" for use after the military conflict between the Revolutionary Armed Forces of Colombia (FARC) and the National Liberation Army (ELN) ended. The management of natural resources occurs mainly at the agroforestry farm level, and there are no formal efforts to manage natural resources. The project proponent, Universidad Industrial de Santander, used three indicators related to regulation and local protection practices: "ecosystem protection"; "maintenance and use of local crop varieties and animal breeds"; and "sustainable management of common resources." The use of these indicators helped community members to evaluate and begin to improve local management systems and issues related to equity, which is of importance to re-inhabited post-conflict areas. Since the state entity in charge of environmental issues in the region (the Regional Autonomous Corporation) restricts the extraction of native timber species from farms where people take care of them, resilience assessment workshop participants felt that they should have greater autonomy in this regard. They showed a growing interest in environmental issues for conservation and to influence communal well-being. The participants also identified serious weaknesses related to cooperation between farmers as well as with other organizations. As a result, they agreed that trust and social capital within the community needed to be strengthened to make real transformation in the landscape.

Under the GEF-Satoyama Project overall, the assessments using the indicators provided a platform for community members, key stakeholders, and project proponents to come together and evaluate the current status of their landscapes and seascapes and share their perceptions with others. They came up with ideas for strengthening governance including official and community regulations, communication, and organizational mechanisms to realize natural resource management in the community. In many cases, assessment participants tried to strengthen collaboration with policy makers to seek better management options. Besides establishing a baseline against which the achievements of the project are measured, resilience assessments provided opportunities to:

- Explore traditional and local knowledge, history, and social and political conditions in the area
- Share strengths and weaknesses of the SEPLS among community members and other stakeholders
- Stimulate dialogue among different community groups and stakeholders that normally do not interact to a significant extent and thus deepen the understanding of differing perceptions toward the landscapes and seascapes
- Understand the needs of local communities
- Strengthen trust between project proponents and the stakeholders

The documentation of the discussion stimulated by the process of attempting to reach consensus scores for the indicators is an asset that endorses participants' thinking, and which will remain as a valuable reference for the future.

6.3.3 A Case Study in Agrobiodiversity from Sierra del Rosario Biosphere Reserve, Cuba

Bioversity International is a global research-for-development organization that puts plant and tree genetic diversity, which nourishes people and sustains the planet, at the heart of its work. Most plant genetic diversity is found in small-scale, traditional agricultural systems largely concentrated in developing countries of the global south, while the productivity and resilience of the world's agriculture depends on a diverse mix of crop varieties, agricultural techniques, farming systems, and traditional and local knowledge. All these attributes can be found in SEPLS, where humans are described as a "keystone species," as they are central to the health of the agroecosystems they have created, and many other species could not survive without their continuous intervention. Bioversity International works with partners on the ground, and through different research projects has tested and used the indicators of resilience in 11 different countries across different agroecosystems, from the highland potato and quinoa plots of Bolivia to the tropical agroforestry systems of Cuba and the rice paddies of China and the Philippines.

The experience with the resilience indicators described here is from Sierra del Rosario Biosphere Reserve in the western part of Cuba. The landscape is home to high levels of agricultural and wild biodiversity and comprises a mosaic of secondary forest patches, home gardens, coffee agroforestry systems, and traditional *conucos*—large gardens or small fields where agriculture is practiced in a traditional way. Cultivated plants in *conucos* came from nearly all regions of the world. Crops from the Central American and Mexican region are most important. The great diversity of different crops as well as the marked variation within most of the cultivated plants demonstrates the importance of the *conuco* as a reservoir for plant genetic resources. Sierra del Rosario is recognized for its rich crop genetic resources of coffee (*Coffea arabica*), maize (*Zea mays*), lima bean (*Phaseolus lunatus*), common bean (*Phaseolus vulgaris*), chili (*Capsicum* species), mango (*Mangifera indica*), plantains and bananas (*Musa* species), and tropical fruits like mamey (*Pouteria sapota*), cherimoya (*Annona reticulata*) and guanabana (*Annona muricata*). Most of the varieties found are traditional.

The application of the resilience indicators in Sierra del Rosario, together with agrobiodiversity and socioeconomic data and information deriving from long-term studies in the same area, highlighted that farmers perceived devastating hurricanes, changes in rainfall patterns, and droughts as the main natural threats to their resilience. However, the landscape's regenerative capacity appears to be relatively high. Thanks in part to interventions based on resilience assessment findings, vegetation patches that were ripped out by two consecutive hurricanes in 2008 showed signs of recovery after only 1 year. Likewise, farmers are adapting to increasingly unpredictable weather and drought by planting more perennial crop species and trees, and by adjusting and changing the timing of agricultural activities. Diversity in land use, crops and crop varieties, as well as smallholder innovation, alternative biodiversity-based livelihoods, and government support are all contributing to resilience to environmental and social changes. The study also highlighted that more work and greater collaboration with the local government, the agricultural cooperatives, and the farmers' association is envisaged to improve small-scale farmers' benefits, and recognition of their contribution to the conservation and production of diverse food items with agroecological methods. Farmers would like to receive more training, by the farmers' association, on new crop varieties and agricultural technologies to improve production, and they see the need for the state agricultural cooperatives to provide better access to specialized markets to add value to biosphere reserve products and agro-ecotourism development.

The assessments carried out by Bioversity International in different countries share some common lessons with the results from Cuba. Community members have come to recognize the usefulness of having, through use of the indicators, a holistic and multidisciplinary approach to their landscapes. Some workshop participants said that they were accustomed to research activities with a narrow focus, and that the resilience assessment was the first time they were involved in such a comprehensive activity. Communities said they felt empowered by expressing their views and aspirations in developing plans for the future of their socio-ecological systems. The sense of ownership, responsibility, and motivation in implementing a plan for their well-being and for their landscape protection was always high among the community members. Communities also felt the need to deepen collaboration among themselves and with extension workers to share and exchange knowledge and experiences. Many of the case studies demonstrated that bringing disparate types of knowledge into conversation can lead to new ways of knowing. In these ways, Bioversity International's experience shows that the indicators have contributed to the understanding and management of complex systems through the lens of worldviews and values of the local communities managing them.

6.3.4 Using the Indicators for Community Self-Diagnosis, Monitoring, and Evaluation in Japan

The indicators were used in surveys conducted as part of a UNU-IAS project to develop a framework and method of monitoring and evaluation for agricultural biodiversity conservation and use in rural villages in Japan, funded by the Ministry of Agriculture, Forestry and Fisheries of Japan. The purpose of the surveys was to capture factors for monitoring including drivers of change and perceived outcomes of conservation activities, whether included in the indicators or not, relevant to the development of monitoring and evaluation methods in Japan. For this purpose, resilience assessment using the indicators was tested as a self-monitoring process for local people that would lead to the creation and revision of action plans. While the indicators had been used mostly in developing countries to date, these surveys were an opportunity to evaluate whether the indicators were also useful in the context of a developed country.

Two assessment workshops were conducted, in the Hiki community of Suzu City, Ishikawa Prefecture in February 2016, and in the Kiyokawa community of Minabe Town, Wakayama Prefecture in August 2016. Both case study sites were chosen as they are located within the Food and Agriculture Organization of the United Nations (FAO) designated Globally Important Agricultural Heritage Systems (GIAHS) sites of "Noto's Satoyama and Satoumi" and "Minabe-Tanabe Ume System," where communities are still utilizing local and traditional knowledge in managing SEPLS for their rural livelihoods today. Agriculture is an important industry for both communities. In the landscapes and seascapes adjacent to Hiki, people are engaged in different kinds of production activities including growing rice and vegetables, charcoal-making, and fisheries; while in the mountainous landscape of Kiyokawa, most are farmers engaged in production of plums (*Prunus mume*), citrus fruits, vegetables, and charcoal.

For each community, a pre-survey was first conducted by questionnaire with 100 local residents prior to a half-day workshop with 15–20 selected participants held to discuss the questionnaire findings (see Table 6.1). This was done to shorten the indicator toolkit's recommended duration of one to two full days for the workshop, as both communities felt that this would be too time-consuming for local residents.

Table 6.1 Overview of the two surveys

	Hiki community, Suzu City, Ishikawa Prefecture	Kiyokawa community, Minabe Town, Wakayama Prefecture
Pre-survey	Date: January 2016 Dissemination: Hand delivery by (sub) community chiefs Collection: Hand delivery to (sub) community chiefs and postal delivery Response rate: 100 sent, 77% response	Date: August 2016 Dissemination: Hand delivery by (sub)community chiefs Collection: Hand delivery to (sub) community chiefs Response rate: 100 sent, 97% response
Workshop	Date: February 2016 Participants: 15 local residents (questionnaire respondents)	Date: October 2016 Participants: 19 local residents (questionnaire respondents)

For the pre-survey questionnaires, questions for each indicator were customized to fit the local context, so that respondents would understand what they were actually being asked. Factors and indicators relevant to the actual circumstances of Japan were then extracted from the questionnaire results and workshop discussions for development of the monitoring and evaluation method. Figure 6.1 summarizes results from the pre-survey questionnaires.

The results of both surveys show that the indicators are useful in providing a broad overview of biodiversity use and conservation status. They were also found to be useful in self-diagnosis to give an overall picture of strengths and weaknesses of the community. Evaluation of biodiversity related to natural resources and agriculture (indicators 1, 2, 3, and 4) tended to be relatively high in both districts, while their evaluation of infrastructure and livelihood (indicators 16, 18, and 19) was lower. Participants understood the importance of evaluating not only ecological and environmental aspects related to conservation and use of biodiversity, but also socioeconomic impacts and how their lives and livelihoods were affected. The assessment captured a wide variety of problems existing in the region and local residents' perceptions of them. Periodic self-diagnosis and self-assessment of this type can be an important monitoring process for local residents to understand changes in their environment and impacts of their activities.

It was also found that the assessment process stimulates conservation in local communities. The workshops turned out to be valuable opportunities for people of different ages, genders, occupations, etc. to gather and discuss. Residents based their assessments on their own daily lives and work experiences, and exchanged views on these during the workshop. This kind of bottom-up, self-assessment method was found useful in: promoting conversation among different age groups and genders, where otherwise in daily life they would not speak directly or share their views; increasing awareness among residents and promoting participation in concrete activities; and helping to clarify roles that each person can and does play in community activities. Thus, the indicators and discussion of results from their implementation were found to be an effective initiating process to promote awareness, establish common understanding, and facilitate planning of activities for resource management.

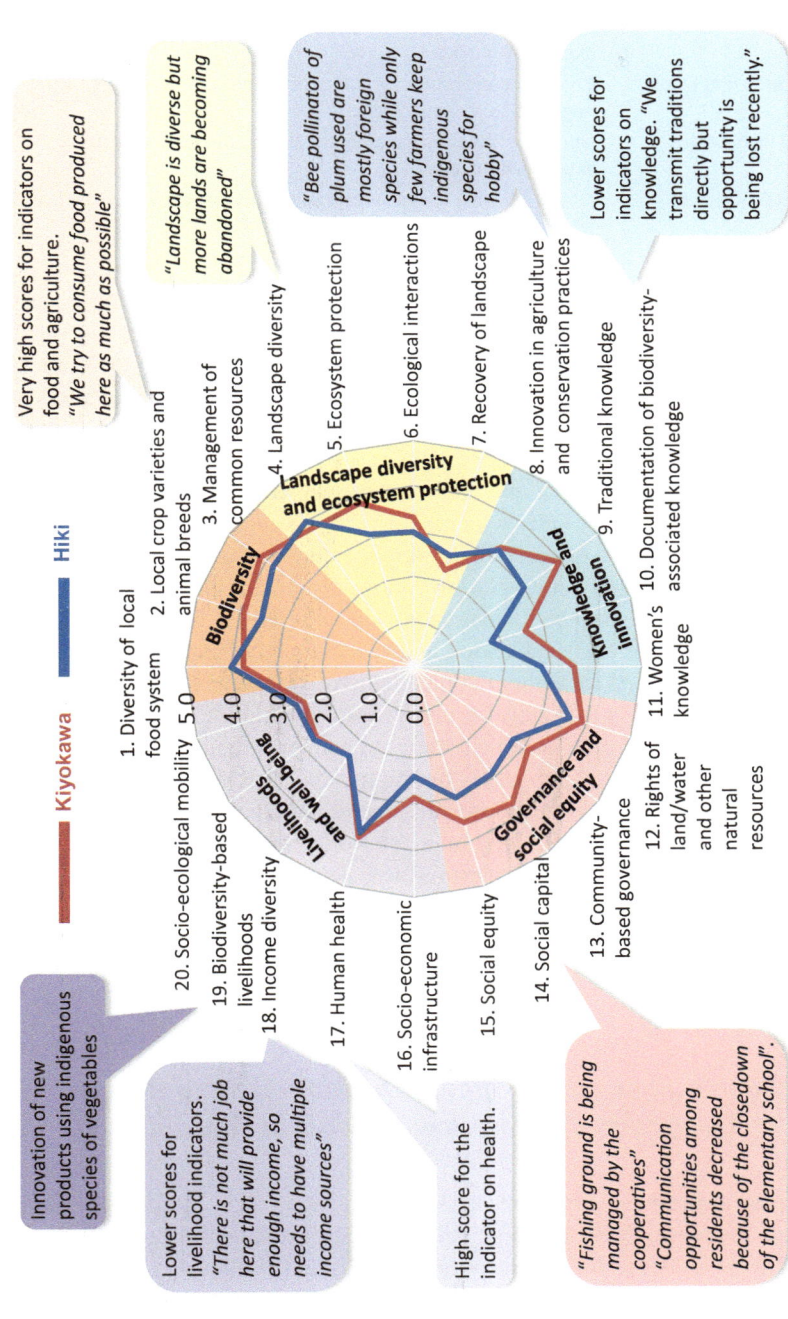

Fig. 6.1 Results from pre-survey questionnaires

Table 6.2 Factors and indicators relevant to circumstances in Japan

Factors	Indicators
Usage of resources	Abandoned cultivation areas on privately owned land, forest management, etc.
Depopulation and aging	Population dynamics including population loss, number of migrants, new farmers, etc.
Absentee landowners	Number of joint management groups, damage from wildlife, extermination, etc.
Spiritual affinity with biodiversity	Nature learning (experience, observation, therapy), arts and culture related to nature, etc.

Besides the 20 indicators, factors and indicators relevant to circumstances in Japan were identified from the pre-survey questionnaires and workshop discussions (see Table 6.2). It was found that not only demographic issues in the population (for example, depopulation) and economic aspects related to biodiversity, but also spiritual and cultural connections were regarded as important factors to motivate interest in conservation of nature. In particular, challenges to the resilience of the communities were mainly due to the lack of workforce, under-use and under-management of resources, and weakening affinity with nature.

For the purposes of the research to develop a monitoring and evaluation method for agricultural biodiversity conservation and utilization activities in rural villages in Japan, the surveys using the indicators proved to be useful for self-diagnosis of current status, strengths and weaknesses, and to capture other factors and challenges that need to be included for a more comprehensive assessment. In particular, this exercise was useful as a first step for conversation within the community and to inspire them to initiate their own activities. Adjustments made to the assessment process, such as conducting pre-survey questionnaires customized to the local context, captured not only more responses in general but also some from those who were not available to commit to long hours of workshop participation. The results also showed that customization of questions was necessary and that additional indicators specific to Japan could be added. With this adjustment, the indicators can be considered to be useful for other rural communities in Japan as well as for wider contexts and purposes.

6.3.5 Supporting Communities in Decision-Making Related to Restoration of Ecosystem Services in Tanzania

Comparative assessment workshops using the indicators were carried out to evaluate community perceptions of diversity in their landscapes and the resources therein, and how these influence local perceptions of risks and resilience, in two villages, Yamba and Kwang'wenda, in Lushoto District, West Usambara, Tanzania in October 2015. These sites were selected because they were among the benchmark sites of CGIAR's Research Program on Climate Change, Agriculture and Food

Security (CCAFS). A combination of different participatory tools was used, including the indicators, as shown in Table 6.3. In each community, 20–25 participants gathered at a central location in the village for the workshop. Each workshop concluded with an in-depth discussion of main problems, root causes and threats identified during the assessment, areas of higher vulnerability and land degradation, and potential solutions, interventions and development organizations.

The following priority actions were proposed in the two villages, aimed at improving the communities' resilience and that of their landscape in the face of socio-ecological and climate changes:

- Restore the ecosystem
- Discourage encroachment on forests, springs, and wetlands through the enforcement of relevant government regulations and policies
- Educate and create awareness on the advantages of diverse foods and landscapes
- Document and preserve community knowledge
- Build capacity of local leaders to lead and groups to network/cooperate
- Support communities to acquire new technologies and innovate
- Diversify sources of income

Six broader themes were identified to categorize the indicators for the scoring exercise (see Table 6.4). Among these, one of the most outstanding differences between Yamba and Kwang'wenda villages is in the first theme, "Landscape/ecosystem diversity and health components." Kwang'wenda village scored 2.7 (average score) in this section against 4.0 (high score) for Yamba village. This was explained by the fact that most forest patches in Kwang'wenda village have been cleared, reducing the tree cover and also the once-numerous springs to only one, causing a general scarcity of water. According to participants, temperatures have risen and rain patterns have been affected, partly attributable to general land degradation and climate change. Crops such as coffee and bananas no longer do well in Kwang'wenda village as it is now too dry and warm for these crops. Yamba village, on the other hand, has suffered much less land degradation. The village has a forest next to it and much of the land still has good tree cover. However, there are reports of forest and spring encroachment by cultivators and illegal timber harvesting, which residents see as a real threat to the ecosystem. The expected future trend follows a similar pattern. Kwang'wenda participants felt that the situation would only improve slightly (average score 2.7) as the state of the ecosystem is bad and the rate of recovery is bound to be slow. In Yamba village, the future looks optimistic (average of 4.2).

The participatory tools used in this study, including landscape mapping, listing of landscape and agrobiodiversity components and main food sources, seasonal calendars, and use of the indicators, also showed a similar association between landscape and species diversity on the one hand and people's perceptions of resilience and risks on the other hand, meaning that people's higher perceptions of landscape and species diversity associate with higher resilience and reduced risks.

Table 6.3 Overview of the assessment workshops

Day	Description	Tool	Estimated time	References
1	Obtain broad overview of the research area. Researchers familiarize with the people, environment, food markets, etc. Identification of focus group participants and recording of personal details such as name, gender, age, village, education, and main occupation.	Transect walk	1 day or more	Geilfus (2008)
	Selection of participants and recording of personal details such as name, gender, age, village, education, and main occupation			
	Making useful plant inventory, demonstrating how to make a plant specimen			Queensland Herbarium (2013)
2	Identification of plant specimens brought by participants			
	Participatory landscape and resource mapping. Understanding of natural and physical resources and distribution. Defining the landscape using local terms. Listing diversity of landscape components	Participatory landscape and resource mapping	2–3 h	IFAD (2009), NOAA (2009), van Oudenhoven et al. (2010)
	Listing agrobiodiversity of landscape components	Agrobiodiversity list	1–2 h	Grum et al. (2008)
	Prioritizing main sources of food and changes over time (past, present)	Historical changes of food and landscape	1–2 h	Catley et al. (2008)
	Historical changes of landscape in timelines	Four cells analysis	1–2 h	Grum et al. (2008)
	Distribution of the food crops. How much are used and by how many people. Availability of food crops and species in different seasons	Seasonal calendar	1–2 h	IFAD (2002), Van de Gevel et al. (2014)
3	Major actors affecting decision-making in management	Venn diagram	½–1 h	Ulrichs et al. (2015), Sontheimer et al. (1999)
	SEPLS indicator scoring evaluation. Understanding of general community perception about resilience in landscapes	Indicators of resilience in SEPLS	2–3 h	UNU-IAS (2014)
	In-depth discussions after indicator scoring, identifying major challenges, potential interventions, and community action plans	Problem analysis, causes, possible interventions	2–4 h	IFAD (2002)

Table 6.4 Mean values of current status and trend scores

Theme	Yamba (mean value of current status/ trend)	Kwang'wenda (mean value of current status/trend)
1. Landscape/ecosystem diversity and health (Indicator 1–4)	4.0/4.2	2.7/2.7
2. Biodiversity including agricultural biodiversity (Indicator 5–8)	3.6/4.1	3.2/4.0
3. Documentation of biodiversity and related local knowledge (Indicator 9, 10)	2.3/2.2	2.6/2.2
4. Landscape resource governance and cooperation (Indicator 13, 14)	1.7/3.9	1.9/4.0
5. Gender knowledge recognition and social equity (Indicator 11, 12 and 15, 16)	4.1/4.2	4.1/4.6
6. Socioeconomic infrastructure, health, and opportunities for income generation (Indicator 17–20)	4.0/3.9	3.3/3.7

Negative perceptions in both score and trend were observed for the themes "Documentation of biodiversity and related local knowledge" (indicators 9 and 10) and "Landscape resource governance and cooperation" (indicators 13 and 14) in both villages. Interventions by institutions, awareness, and institutional collaboration (indicators 13 and 14) were observed to have a great influence on perceived trends and optimism that there will be better resource management in the future. Knowledge documentation, gender, and social equity have the least influence currently and also in the future, as shown by "Documentation of biodiversity and related local knowledge." Knowledge documentation was considered by the community as inconsequential for their livelihoods and resilience, but some participants pointed out that local knowledge was still being passed from parents to their children by traditional means such as storytelling and physical interactions. Gender issues and social equity were generally ranked high, with the future remaining bright. Most men felt there was equitable sharing and equal rights with respect to resource access and sharing, and that women's knowledge was respected, but women saw room for improvement.

Differences between Yamba and Kwang'wenda villages in their perceptions of landscape resilience and levels of optimism were clear. Yamba has a greater landscape diversity, higher level of food diversity, and higher perceived resilience and optimism than Kwang'wenda, as it is close to forests, rivers, and springs; and hence more agroecological zones, habitats, and niches that support species diversity. Yamba also has a greater choice of resources, better infrastructure, and more income opportunities. Community members, however, highlighted encroachment on forests and springs, poor leadership, lack of enforcement of environmental protection laws, and lack of locally coordinated collective actions to manage these resources. Kwang'wenda, on the other hand, is located on a hilly landscape with many steep-sided hills. Farming is done on the hilltops, on the slopes, and in the valleys. There is a perception of serious risks related to soil erosion and poor productivity, and a need for conservation agriculture due to generalized deforestation depleting their natural resources. Needs for good leadership, collective action, networking, and institutions working together toward common goals were strongly indicated.

The combination of participatory assessment tools and processes used for this study highlighted people's perceptions of resilience, its major determinants, and available options for improvement. Establishing an innovative and community-based multi-level coordinating body or committee is a key element for adoption of solutions proposed by participants. The collective information identified through a combination of different participatory tools in Table 6.3 can be used as an extension guide to: facilitate the community to network with different stakeholders to discuss and share knowledge; facilitate training on how to raise funds and write proposals to support their initiatives; facilitate training workshops and seminars for government officials; and mobilize support for improvement of social infrastructure including roads, health facilities, irrigation systems, wells, or boreholes. It can also contribute to building a multi-lateral strategy for restoration of degraded ecosystems, as well as to monitor effectiveness of interventions.

6.4   Discussion

The diversity—in terms of type, approach, purpose, geography, and others—among the examples presented here is an indication of the difficulties involved in assessing resilience, as resilience itself is expressed in widely varying ways by communities in different socio-ecological contexts. There are a multitude of factors—ecological, economic, well-being, or governance-related—and challenges in knowledge transfer and management of resources and ecosystems that can affect resilience positively or negatively in different production landscapes and seascapes. Due to the sector-based approach taken in most project and intervention planning, people have tended to consider these factors separately, although on the ground they are part of a system with complex, interconnected cause–effect relationships. The value of an indicators approach like the one described in this chapter is its ability to simultaneously unravel components of SEPLS into factors that can be realistically assessed, while not losing sight of the fact that they make up an interconnected system. This thereby encourages just and sustainable use and management of resources, better planning of livelihood activities, and cooperative partnership building among various stakeholders.

Nearly 10 years of experience with developing, testing, and implementing the indicators of resilience in SEPLS has produced a great deal of knowledge on this kind of indicators approach. Among the major benefits found in using the indicators in all of these cases is their value in organizing local peoples' perceptions by helping stakeholders to understand and identify elements of socio-ecological resilience, challenges to resilience, and potential strategies to reach and sustain ecological integrity and human well-being. With this base of improved understanding, the indicators can then serve as a tool for monitoring achievements and progress toward improved resilience, as seen in the experience of the COMDEKS program. At the same time, use of the indicators can help communities identify gaps in their understanding of system complexity and dynamics, and thereby identify opportunities to leverage synergies, as seen in the Cuban case and others.

A unique aspect of this particular indicators approach is that its inclusion of temporal trends and timelines has helped communities' understanding of historical resource-use and management practices and consumption effects. Likewise, the inclusion of community-based mapping exercises into resilience assessment workshops has identified resource-use patterns and drivers of change that affect the current situation as well as positive or negative effects of potential interventions. The resulting more comprehensive understanding of socio-ecological interactions in the landscape or seascape over time has promoted greater appreciation for and commitment to revitalization of traditional and local knowledge and practices, and consequently empowerment of local communities, for example in the cases from the "GEF-Satoyama Project." This empowerment extended to groups within the landscape or seascape by incorporating perspectives, knowledge systems, and worldviews of vulnerable and under-served segments of the community, encouraging planning by internal stakeholders with high transformational potential, as in the case from Japan and those from Bioversity International.

Possibly the most prominent benefit found in these experiences using the indicators is their value as a convening tool, bringing together multiple stakeholders in a landscape or seascape. Using the toolkit has helped with stakeholder engagement by bringing together groups that did not previously communicate much in the landscapes and seascapes presented here, allowing them to work together toward building shared understanding and goals between local communities, governments, and other stakeholders to promote participatory management. This includes a more comprehensive understanding of trade-offs between different priorities of different actors and actions, and the types of conflicts that can arise, allowing more informed efforts to overcome undesirable outcomes from governance and management actions, as in the case from Colombia.

Overall, this indicators approach has been found to be a good tool for the purposes above, and others including its value in contributing to scientific research, as seen in the case from Tanzania, and to encourage conservation, as in the Japan case, by helping local people to understand the links between ecosystem services and their own livelihoods. That said, some points for further consideration when using the indicators were identified in these cases. For example, some found that the indicators may be more effective if they are adapted for local contexts by adding or modifying some indicators as appropriate, or combined with other tools as in the Tanzanian case. This is one result of the indicators' use of a community-based approach, which, as indicated above, allows for and also requires a variety of ecological, socioeconomic, and other factors to be considered in an integrated approach to assess resilience at the community level (Antwi et al. 2014).

Most importantly, users stressed that, due to their subjective nature and widely varying socio-ecological circumstances, these indicators cannot be used for comparison of different landscapes and seascapes, as they rely on community members' individual perceptions rather than third-party evaluation. This locally specific nature of the assessment process was seen as a strength as well as a disadvantage by users, due to the fact that the factors contributing to resilience themselves have been found to be highly context-specific (Saito et al. 2018) and therefore are difficult to assess using common indicators that may not be relevant to local circumstances. In other words, a locally specific integrated assessment model can help understand resilience better, but only at the expense of generalizability across different communities. While the indicators cannot provide comparison between geographical areas, they can be used to capture changes within a community over time through repeated assessment, particularly if the same group of participants can do the assessments. This can be valuable for adaptive management of the landscape or seascape, as highlighted in Sect. 6.3.1 above.

Work already done with the indicators as described above has indicated some directions for future work. For one, this approach's potential role in monitoring and evaluation of project results was identified as an area that has not yet been sufficiently explored, and where there seems to be some future potential. Another is further investigation of the use of the indicators in developing- versus developed-country contexts. To date, the assessments included in this chapter from Japan are the only ones that were done in countries classified as developed countries. Finally, users stress that there is further work to do in scaling-up the indicators and assessment activities to capture the implications of local-level resilience for larger governance and ecological scales. If this avenue is pursued, "Indicators of Resilience in Socio-ecological Production Landscapes and Seascapes" could have further potential relevance for national and international conservation and development targets, such as the Aichi Biodiversity Targets of the Convention on Biological Diversity's Strategic Plan for Biodiversity 2011–2020, and the United Nations' Sustainable Development Goals.

References

Antwi E, Otsuki K, Saito O, Obeng FK, Awere Gyekye K, Boakye-Danquah J, Boafo Y, Kusakari Y, Yiran G, Owusu A, Asubonteng K, Dzivenu T, Avornyo V, Abagale F, Jasaw G, Lolig V, Ganiyu S, Donkoh S, Yeboah R, Takeuchi K (2014) Developing a community-based resilience assessment model with reference to Northern Ghana. J Integr Disaster Risk Manage 4. https://doi.org/10.5595/idrim.2014.0066
Bergamini N, Blasiak R, Eyzaguirre P, Ichikawa K, Mijatovic D, Nakao F, Subramanian SM (2013) Indicators of resilience in Socio-ecological Production Landscapes (SEPLs). United Nations University Institute for the Advanced Study of Sustainability, Tokyo
Catley A, Burns J, Abebe D, Suji O (2008) Participatory impact assessment. A guide for practitioners. Feinstein International Center, Boston

Geilfus F (2008) 80 tools for participatory development: appraisal, planning, follow-up and evaluation. IICA, San Jose

Grum M, Gyasi EA, Osei C, Kranjac-Berisavljevic G (2008) Evaluation of best practices for landrace conservation: farmer evaluation. Bioversity International, Rome

International Fund for Agricultural Development (2009) Good practices in participatory mapping. International Fund for Agricultural Development, Rome

International Fund for Agricultural Development Gender Strengthening Programme for Eastern and Southern Africa Division (2002) Toolkit for practitioners: gender and poverty targeting in market linkage operations. International Fund for Agricultural Development. https://www.ifad.org/documents/10180/a0a52e39-ec33-4118-b8c5-aa99b5a23be3.

NOAA Office for Coastal Management (2009) Stakeholder Engagement strategies for participatory mapping. https://coast.noaa.gov/data/digitalcoast/pdf/participatory-mapping.pdf.

Queensland Herbarium (2013) Collection and preserving plant specimens, a manual. Department of Science, Information Technology, Innovation and the Arts, Brisbane

Saito O, Boafo Y, Jasaw G, Antwi E, Shoyama K, Kranjac-Berisavljevic G, Yeboah R, Obeng F, Gyasi E, Takeuchi K (2018) The Ghana model for resilience enhancement in Semiarid Ghana: conceptualization and social implementation. In: Saito O, Kranjac-Berisavljevic G, Takeuchi K, Gyasi E (eds) Building integrated resilience strategy against climate and ecosystem changes for Sub-Saharan Africa (Series: Science for Sustainable Societies). Springer, Japan, 343p

Sontheimer S, Callens K, Seiffert B (1999) Conducting a PRA training and modifying PRA tools to your needs. An example from a Participatory Household Food Security and Nutrition Project in Ethiopia. http://www.fao.org/docrep/003/x5996e/x5996e00.HTM. Accessed 1 June 2018

Stockholm Resilience Center (2014) What is resilience? An introduction to socio-ecological research. http://www.stockholmresilience.org/download/18.10119fc11455d3c557d6d21/1459560242299/SU_SRC_whatisresilience_sidaApril2014.pdf. Accessed 18 Nov 2018

Ulrichs M, Cannon T, van Etten J, Morimoto Y, Yumbya J, Kongola E, Said S, van de Gevel J, Newsham A, Marshall M, Kabululu S, Kiambi D, Nyamongo D, Fadda C (2015) Assessing climate change vulnerability and its effects on food security: Testing a new toolkit in Tanzania. Working Paper No. 91. CGIAR Research Program on Climate Change, Agriculture and Food Security, Copenhagen

United Nations Development Programme (2018) Assessing landscape resilience: best practices and lessons learned from the COMDEKS Programme. UNDP, New York

UNU-IAS (2010) Satoyama-Satoumi ecosystems and human well-being: socio-ecological production landscapes of Japan – summary for decision makers. United Nations University, Tokyo

UNU-IAS, Bioversity International, IGES and UNDP (2014) Toolkit for the indicators of resilience in socio-ecological production landscapes and seascapes. United Nations University Institute for the Advanced Study of Sustainability, Tokyo

van de Gevel J, Bijdevaate M, Mwenda P, Morimoto Y, Fadda C (2014) Guiding focus group discussions on varietal diversification and adaptation to climate change in East Africa. Bioversity International, Rome

van Oudenhoven F, Mijatovic D, Eyzaguirre PB (2010) Bridging managed and natural landscapes. The role of traditional agriculture in maintaining the diversity and resilience of social-ecological systems. In: Bélair C, Ichikawa K, Wong BYL, Mulongoy KJ (eds) Sustainable use of biological diversity in socio-ecological production landscapes: background to the 'Satoyama' Initiative for the Benefit of Biodiversity and Human Well-Being. CBD Technical Series 52. Secretariat of the Convention on Biological Diversity, Montreal, pp 8–21

van Oudenhoven F, Mijatović D, Eyzaguirre P (2011) Social-ecological indicators of resilience in agrarian and natural landscapes. Manage Environ Qual 22(2):154–173. https://doi.org/10.1108/14777831111113356

Chapter 7
Place-Based Solutions for Conservation and Restoration of Social-Ecological Production Landscapes and Seascapes in Asia

Raffaela Kozar, Elson Galang, Jyoti Sedhain, Alvie Alip, Suneetha M Subramanian, and Osamu Saito

Abstract The relevance of traditional land-use systems in Asia is under threat from externally influenced drivers such as the use of modern agricultural technologies, urbanization, rapid industrialization, overexploitation, and underutilization. The impacts of these changes in land use are contributing to a loss of biodiversity and ecosystem services (BES) in social-ecological production landscapes and seascapes (SEPLS). Societal actors operating from multiple scales create and implement place-based solutions in SEPLS in response to landscape-specific challenges and opportunities for achieving biodiversity conservation and sustainable development. This study aims to identify and demonstrate the abundance of place-based solutions for solving challenges to sustainable use and management of natural resources in SEPLS, and to better inform the existing suite of conservation and restoration solutions. We review a set of 88 case studies from The International Partnership for the Satoyama Initiative (IPSI) in the South, East and Southeast Asian regions using a societal-based solution scanning approach to systematically identify these solutions for conservation and restoration at local scales and to categorize them by solution type. Societal actors demonstrate preferences for solution types to reversing the loss

R. Kozar · E. Galang · J. Sedhain · A. Alip
United Nations University Institute for the Advanced Study of Sustainability (UNU-IAS), Tokyo, Japan

S. M. Subramanian
United Nations University International Institute for Global Health (UNU-IIGH), Cheras, Kuala Lumpur, Malaysia

O. Saito (✉)
United Nations University Institute for the Advanced Study of Sustainability (UNU-IAS), Shibuya, Tokyo, Japan

Institute for Global Environmental Strategies (IGES), Hayama, Kanagawa, Japan

Institute for Future Initiatives (IFI), The University of Tokyo, Bunkyo, Tokyo, Japan
e-mail: saito@unu.edu

© The Author(s) 2020
O. Saito et al. (eds.), *Managing Socio-ecological Production Landscapes and Seascapes for Sustainable Communities in Asia*, Science for Sustainable Societies, https://doi.org/10.1007/978-981-15-1133-2_7

117

of BES in SEPLS while embracing a mix of all solution types across ecosystems. Institutional and governance solutions are the most common type across Asia. Technological solutions are preferred in East Asia, while knowledge and cognitive solutions are preferred in Southeast Asia. Economic and incentive-based solutions are found most often in South Asia as livelihood investments for local residents, and to balance trade-offs among food production and biodiversity conservation. Sharing the knowledge of various place-based solution types in different social-ecological contexts helps improve more purposeful and deliberate design of SEPLS for multiple benefits.

Keywords Place-based · Solution scanning · Biodiversity conservation · SEPLS · Sustainable development

7.1 Introduction

7.1.1 An Accelerating Loss of Biodiversity and Ecosystem Services in Asia

Communities across Asia are facing unprecedented threats to traditional natural resource-based livelihoods in managed agricultural landscapes (van Oudenhoven et al. 2010). The loss of biodiversity and ecosystem services (BES) from land-use change is affecting the systems communities have relied on for the sustainable use and management of resources that provide for their livelihoods. This changing relevance of traditional land-use systems in Asia is driven by a number of externally influenced drivers such as the use of modern agricultural technologies, urbanization, rapid industrialization, overexploitation and underutilization. We introduce some of the ways each of these drivers have upset the balance of resource use in turn.

First, the use of modern agricultural technologies has changed the balance of species in some ecosystems (Kumar and Takeuchi 2009; Plieninger et al. 2014; Akça et al. 2015; Katayama et al. 2015; Aadrean 2017). In Taiwan's Shungxi River Valley for instance, farmers' use of chemical fertilizers and pesticides has put great pressure on local aquatic species and other freshwater-based ecosystem services (Yun-Ju et al. 2015). Increasing urban sprawl and changing consumer demands are driving shifts in ecosystem composition, the loss of agrobiodiversity, and the services provided to humans across communities in the region (Kumar and Takeuchi 2009; Knight 2010; Kohsaka et al. 2013; Plieninger et al. 2014; Sakurai et al. 2016; Yu et al. 2016). In Pakistan's Jhelum River Basin, urbanization has been credited with the reduction of wildlife species, shifts of indigenous plant species to non-native species, and increased contamination of the river (Khan et al. 2017).

Industrialization in South Korea was key to the country's economic growth in the 1960s, but came in part at the expense of the country's traditional rural agricultural production landscapes known as "maeuls" (UNU-IAS 2012a). Rapid industrialization is now a factor in the loss of rural production landscapes and the diversity of crops grown within them in multiple countries across Asia (Knight 2010; Shimada 2015; Tomita et al. 2015). Overexploitation of resources from increasing population pressures and multiple demands by actors at different levels, together with poverty and the exacerbating effects of climate change, is a continuing threat to BES in traditional land-use systems (Shimada 2015; Takeuchi et al. 2016). In Cambodia's Chumkiri District, anarchical forest exploitation by both insiders and outsiders has turned the area's semi-jungle forest into a degraded forest, significantly impacting the local people who have long been dependent on various forest-based ecosystem services (Marady et al. 2011).

Finally, underutilization is a relatively unique phenomenon to countries where populations are decreasing or aging and driving the loss of BES through the transformation of abandoned lands to new ecosystems (Putra and Nakamura 2009; Kieninger et al. 2011; Morimoto 2011; Tsuchiya et al. 2013; Plieninger et al. 2014; Queroiz et al. 2014; Katayama et al. 2015; Shimada 2015; Li and Li 2016; Osawa et al. 2016; Takeuchi et al. 2016). For instance, in Japan's Toyooka City, underutilization in the form of abandonment of farmlands served as a major factor in the disappearance of oriental white storks, an important fauna in maintaining the area's wetland ecosystem (Toyooka City 2012).

A recent regional assessment of the status of BES in Asia and the Pacific found that the interaction of these factors and others are accelerating the rate of loss of BES across Asia in ways that are threatening livelihoods and food security, but found that management choices can alter this trajectory (IPBES 2018).

7.1.2 Community-Based Sustainable Use and Management of Resources

Communities have practiced varying sustainable resource management approaches in places where multiple ecosystem services for human well-being are derived from patchwork land uses and ecosystems in mosaics with human settlements, known as social-ecological production landscapes or seascapes (SEPLS) (Bélair et al. 2010; Hashimoto et al. 2015). One such approach to managing SEPLS for multiple benefits is embodied in the Japanese concept of "Satoyama," or the balance of society and nature in harmony. Many experiences with the Satoyama tradition of sustainable use and the conservation of SEPLS have been documented (See for instance the volumes of: Subramanian et al. 2015, 2016, 2017, 2018; Okayasu and Matsumoto 2013).

However, the traditional Satoyama approach, which customarily provided multiple ecosystem services and benefits to humans, no longer suffices in practice to deliver the same level, mix or synergies among ecosystem services due to the externally influenced drivers of an accelerating loss of BES that are changing the patterns and feedbacks among land use and users at different scales (Takeuchi et al. 2016). In the examples in the preceding section, not only did a loss of BES impact people's livelihoods and well-being, but re-shaped the overall dynamics of diverse and complex social-ecological systems in communities across Asia. Impacts of a loss of BES on water availability and quality, crop productivity, health, nutrition, and food security are causing SEPLS to no longer serve as reliable sources of income and support for the well-being of local residents.

7.1.3   The New Challenges to Sustainable Use and Management of SEPLS

The drivers of a loss of BES and their interactions are complex and span multiple scales. They also introduce new demands on the ecosystems in the landscape, while the stakeholders making such demands are increasingly located at multiple social-ecological scales. Therefore, local residents together with other societal actors spanning social-ecological scales and seeking to address the same challenge need to work together on the solutions. In the rest of this chapter, we use the term "place-based" solutions to refer to these solutions that are developed for use in a particular SEPLS by societal actors at different social-ecological scales, and that have a wide range of values shaping their preferences for selection of solutions.

Addressing this challenge of an accelerating loss of BES and revitalizing SEPLS in ways that will balance society and nature in harmony requires sustainable use and management approaches that function across multiple scales and that take into account the feedbacks in social-ecological systems (Takeuchi et al. 2016). Therefore, increasingly, approaches to sustainable use and management of SEPLS need to address the multiple externally influenced drivers that are challenging traditional land-use patterns, and also account for the wider range of values from different users from multiple scales, while still meeting the livelihood and well-being of local populations (Gu and Subramanian 2014; Havas et al. 2016; Duraiappah et al. 2014; IPBES 2015; Bohnet and Beilin 2015).

The 2018 IPBES regional assessment for BES in Asia finds that the types of management approaches that can best help address the links among drivers that are accelerating a loss of BES are those that are based on community approaches to sustainable use and management and that link multiple stakeholders and levels through collaborative decision-making processes (IPBES 2018). These types of approaches put communities at the center of defining the priority values of land use and decision-making in the landscape while linking the management functions and jurisdictions over resource governance across scales. These approaches include various forms of a landscape-scale approach to integrated management of the benefits from BES in SEPLS. Not only should these management approaches address ecosystem services today, but also the needed ecosystem services in the future for resilient ecosystem function and services across scales (IPBES 2015; Oliver et al. 2015).

New sustainable use and management approaches require a ready suite of implementable solutions for the revitalization and conservation and restoration of BES in SEPLS that can account for these changing dynamics and future needs, while working across multiple actors and scales. In addition, the synergies and trade-offs among conservation and restoration solutions across the mosaic ecosystem character and multiple scales of SEPLS have to outreach the expected impacts of climate change, which are likely to impede progress in development goals in South and Southeast Asia and exacerbate biodiversity loss in the hotspots of Asia if management aims that can meet these criteria are not met (Springmann et al. 2016).

Devising new sustainable use and management approaches that communities can embrace and implement requires understanding how collaboration can deliver multiple solutions that can balance multiple functions in SEPLS and that benefit multiple stakeholders (Cockburn et al. 2018; Freeman et al. 2015). The next section looks at how readily deployable conservation and restoration solutions are to meet the challenges to sustainable use and management of SEPLS.

7.1.4 Conservation and Restoration Solutions for SEPLS

Conservation and restoration solutions in SEPLS redress the loss of BES from land-use change. Extensive work has been done to catalog these conservation solutions based on expert knowledge (Sutherland et al. 2017; see for instance Dicks et al. 2016). Yet the wealth of solutions held by societal actors have not been given the same weight as expert opinion in the scientific evidence base for effective solutions. To meet the most pressing sustainability challenges in biodiversity and climate, science and evidence-based solutions are not enough.

Societal actors are those that form in response to a given challenge at a given point in time or over a period of time to provide solutions. Communities across Asia are working together with other societal actors from across scales to implement place-based solutions in response to the challenges and opportunities for achieving biodiversity conservation and sustainable development in SEPLS. A better understanding of these place-based solutions and how they can be selected and utilized for sustainable use and management can help expand the current state of knowledge of conservation and restoration solutions with the best available knowledge reflecting multiple values (Jacobs et al. 2016). Integrating multiple knowledges enriches solutions and leads to the potential for place-based solutions not previously under consideration or those that reflect the dynamics of changing landscapes, while increasing acceptability among societal actors with different values.

This study aims to identify and demonstrate the abundance of place-based solutions for solving challenges to sustainable use and management of natural resources in SEPLS, and to better inform the existing suite of conservation and restoration solutions in the scientific literature with the experiences of societal actors.

The rest of this chapter is organized as follows. Section 7.2 describes the methodology for a societal-based approach to systematically identifying solutions. The third section describes the experiences of local residents in navigating solutions with other societal actors at different scales and across ecosystems. Section 7.4 explores the socio-environmental contexts of different solutions and discusses their implications for informing sustainable use and management approaches. The final section draws some conclusions and suggests areas that need further attention in advancing conservation and restoration solutions that deliver sustainable use and management of SEPLS in the context of pressing sustainability challenges in biodiversity and climate.

7.2 Methodology

7.2.1 A Societal-Based Approach to Solution Scanning

Solution scanning is a systematic process of making an inventory of all possible responses to a problem prior to weighing the feasibility and merit of each solution for use in a particular setting (Sutherland et al. 2014). It's been used in environmental and sustainability research literatures to identify solutions for maintaining ecosystem services (Sutherland et al. 2014), agroforestry-based solutions for climate mitigation and adaptation (Hernández-Morcillo et al. 2018) and to scan for existing food network models as a solution type in cultural landscapes in Europe and Asia (Plieninger et al. 2018). The three-part cycle starts with problem identification (horizon scan), then secondly the solution scan, and third the filtering process, which is how solutions can be weighed and selected for their effectiveness in a particular context (Sutherland et al. 2014).

In a recent review of multi-level networks and sustainability solutions, we proposed a societal-based solution scanning approach (Kozar et al. 2019). Transdisciplinary methods that engage multiple types of knowledge, when used with a sustainability science framework focused on societally relevant problems, can help address the questions related to which solutions for a loss of BES can be most effective in managing SEPLS (Pascual et al. 2017). These solutions should be selected for paths that preserve the multiple benefits and ecosystem services borne by traditional management systems, while considering current and future needs through multiple stakeholder values at multiple levels and across the mosaic ecosystem character of SEPLS.

7.2.2 *Methods*

This chapter presents place-based solutions for conservation and restoration of SEPLS in the Asian region. These are presented: (1) in different sub-regions of Asia at the local scale, (2) by ecosystem, and (3) by how they are connected to other scales through multi-level governance by coalitions of societal actors. We drew upon the experiences of the International Partnership for the Satoyama Initiative (IPSI) in identifying and implementing solutions.[1] IPSI seeks to conserve and revitalize SEPLS through rejuvenation of the Satoyama approach in social-ecological systems that face current environmental challenges from land-use changes. IPSI does this while sustainably supporting the livelihoods and well-being of local communities through revitalized, adaptive and innovative production and management systems that are evolved from cultural practices and indigenous and community knowledge (Gu and Subramanian 2014; Takeuchi et al. 2016; Berglund et al. 2014).

IPSI is a coalition of societal actors that have agreed to share knowledge and collaborate to improve the management of SEPLS in response to evolving threats, including the loss of BES caused by the interactions of a multitude of externally influenced drivers. In Step 1, the horizon scan, the problem we chose to focus on is the one identified by the societal actors of the IPSI network, which is a loss of biodiversity and ecosystem services. The network has aimed to share place-based solutions to this problem over the past decade.

For step 2, the solution scan, we selected case studies from all publicly available IPSI member cases up to March of 2018. Cases are from 2009 to 2017 and from four primary sources: (1) an online case study database hosted by IPSI (UNU-IAS 2018); (2) a publication by the Satoyama Initiative on Asian production landscapes (UNU-IAS 2012b); (3) publications from the Communities in Action for Landscape Resilience and Sustainability—The COMDEKS program, produced through a collaborative activity of IPSI (UNDP 2014a, 2016); and (4) the flagship series of the Satoyama Initiative, its annual thematic review (Subramanian et al. 2015, 2016, 2017).

We used a set of categorical variables collected in a Microsoft Excel based data instrument. Data collection instruments were created and refined through consultations and pre-testing. A data definition and collection guide was developed to inform the data collection process, and included variables and their range of values, definition, and collection instructions. Quality assurance was controlled through three rounds of pre-testing the data collection sheet by the research team made up of the authors, whereby after each round the range of values and definitions were revised by the shared understanding among team members. Another quality assurance mea-

---

[1] The International Satoyama Initiative (ISI) was jointly initiated by the Ministry of the Environment of Japan (MOEJ) and the United Nations University Institute for the Advanced Study of Sustainability (UNU-IAS). On 19 October 2010, during the tenth meeting of the Conference of the Parties to the Convention on Biological Diversity (CBD COP 10), the International Partnership for the Satoyama Initiative (IPSI) was established to promote the activities identified by ISI and to share relevant information and experiences.

sure included the use of a pre-defined set of values for many of the variables in the data collection sheet to reduce error in data entry, and the aforementioned guide to definitions and response types for the remaining open entry cells. Finally, after data entry, responses were standardized with consistent terms, and various consistency checks performed.

Types of data collected included case study information (publication year, location, scale, institutional author); socioeconomic and biophysical information (sectors, stakeholders, institutions, livelihoods, threats, ecosystems, and ecosystem products); program information (goals, institutional and legal environment, outcomes, knowledge mechanisms); and solutions (solution, solution type). We defined a solution as any activity, intervention, innovation, practice, strategy, or policy that has been proposed or applied in the case study area to address the given problem.

Data for 91 cases was recorded. A minimum criteria was set for each case study to include at least one solution and at least one value for all data categories in the socioeconomic and biophysical section. One case was excluded that did not meet the minimum criteria. A data cleaning protocol was applied to the remaining 90 cases. During the data cleaning, data from case studies in the same location and with the same institutional author were merged in 2 cases, resulting in a total of 88 cases. The final number of case studies and their geographical locations are shown in Table 7.1. A total of 23 cases are located in South Asia, 29 cases in Southeast Asia, and 36 cases in East Asia, representing 18 countries in these regions. Fifty-two local-scale cases were identified by reviewing these 88 case studies, and those countries with local-scale cases are indicated in bold script in Table 7.1.

For step 3, the filtering process for selection of solutions to apply in a particular context, we did not discuss or evaluate solutions based on their effectiveness for a particular place. Rather, we applied a framework to filter the solutions by solution type (adapted from the Millennium Ecosystem Assessment 2005) in order to understand which type of solutions societal actors might prioritize in different social-ecological contexts.

Table 7.1 Location and number of study cases from the International Partnership for the Satoyama Initiative (IPSI)

Southeast Asia	South Asia	East Asia
Cambodia (8)	Bangladesh (1)	**China (15)**
Indonesia (6)	**Bhutan (2)**	**Japan (18)**
Lao People's Democratic Republic (1)	**India (9)**	Mongolia (2)
Myanmar (1)	Iran (1)	**South Korea (1)**
Philippines (5)	**Nepal (7)**	Total 36
Vietnam (4)	**Pakistan (2)**	
Thailand (4)	Sri Lanka (1)	
Total 29	Total 23	

Bold font indicates countries with local-scale case studies. Numbers in parenthesis indicate the number of case studies included in the total 88 cases of the review

A workshop was held with the research team. Using a consensus process, the solutions were categorized and a further typology of 25 sub-categories was developed. Solutions for the 88 case studies were categorized according to the following 5 categories: institutional and governance solutions; economic and incentive-based solutions; social, cultural, and behavioral solutions; knowledge and cognitive solutions; and technological solutions (Table 7.2).

The conventional step 3 filtering process of the solution scan method aims to determine which solution should be applied in a certain place and context based on some agreed expert criteria for effectiveness such as budget, feasibility, and time. In a societal-based solution scanning approach, which solutions to apply in a given SEPLS should be determined in a place-specific and transdisciplinary manner, including knowledge from societal actors on the preferred benefits and trade-offs in the process.

Ecosystems in the case studies were recorded according to the ten classification types of the Millennium Ecosystem Assessment (marine, coastal, inland water, forest, dryland, island, mountain, polar, cultivated, urban) (MA 2005). Up to four ecosystems per case study were recorded to capture the mosaic characteristic of SEPLS.[2] No cases were located in polar and marine ecosystems in the overall set of 88 cases.

In this chapter, the results discussed present those solutions identified at local scales. These scales are at village, sub-municipality, and local government levels on social and administrative scales, and water bodies (river, lake) or watershed scales at ecological scales. They include both solutions already existing and implemented in the case study areas and those that are proposed as solutions to the given challenge. We examined a sub-set of 52 case studies to answer our question on sub-regional experiences, the societal actors engaged in navigating solutions and the ecosystems targeted through local-level solutions.

7.3 Experiences, Actors, and Ecosystems in Navigating Place-Based Solutions

Solutions at the local scale represent 283 solutions in 52 case studies, or 58% of the total 485 solutions identified in 88 case studies (Kozar et al. 2019). When comparing the distribution in choice of solutions by solution type between the local scale and higher scales (other administrative or ecological scales such as national level or coastline), institutional and governance solution types are selected more often at higher scales, 34–26% (Table 7.3). Technological solutions are more often selected at local scales, 21–12%. Solution types are more evenly distributed at local scales, compared to higher scales where there is a 15% higher rate of selection for institu-

---

[2] Analysis of findings per ecosystem in this chapter is in regard to the "first" or main ecosystem representing the case study area.

Table 7.2 Categories and sub-categories of solutions

Solution type	Description	Sub-categories
Institutional and governance	Solutions that enhance benefits while conserving resources by addressing weak or insufficient institutional and management systems and with coordinated responses at multiple scales that consider regulation of ecosystem services in the long term	• Organizational development and institutional strengthening
		• Integrated management approaches
		• Regulations, policies, or frameworks
		• Inclusion
		• Financing
		• Enabling conditions
Economic and incentive-based	Solutions that address market failures and misalignment through market-based approaches along with improved value-chains and consumer preferences	• Taxes and user fees
		• Subsidies, payments, and rewards
		• Improved value-chains
		• Consumer preferences
		• Trade systems
		• Livelihood
		• Market access
Social, cultural, and behavioral	Solutions that reduce demand or consumption or address the lack of political and economic power of some groups who are particularly dependent on ecosystem services or harmed by their degradation through demand-side responses	• Formal and nonformal education
		• Awareness creation
		• Cultural practices
		• Access to social services
Knowledge and cognitive	Solutions that address insufficient knowledge or the poor use of existing knowledge concerning ecosystem services, address information gaps and incorporate other forms of knowledge and information	• Knowledge integration
		• Knowledge gaps
		• Knowledge capacities
		• Knowledge systems
Technological	Solutions that reduce the harmful impacts of various drivers of ecosystem change as well as underinvestment in the development and diffusion of technologies, or that could increase the efficiency of resource use or ecosystems	• Agroecological practices
		• Ecological restoration or conservation practices
		• Energy technologies or investments
		• Green and resilient infrastructure

Adapted from MA (2005)

Table 7.3 Solutions by type at local and other scales, and overall, by number and percent

	Local (%)	Other administrative or ecological scale (%)	Overall results (%)
Economic and incentive-based	57 (20)	36 (18)	93 (19)
Institutional and governance	74 (26)	68 (34)	142 (29)
Knowledge and cognitive	42 (15)	34 (17)	76 (16)
Social, cultural, and behavioral	50 (18)	39 (19)	89 (18)
Technological	60 (21)	25 (12)	85 (18)
Total	283 (100)	202 (100)	485 (100)

tional and governance solutions (34%) relative to the rate of selection of the next most often selected solution type, social, cultural, and behavioral (19%). The distribution among the 5 solution types at local scales compared with the overall distribution among 485 solutions does not show a demonstrable difference.

We now turn to look at how these solution types break down in experiences per sub-region, among societal actors, and in different ecosystems.

7.3.1 Sub-regional Experiences and Variation of Place-Based Solutions

Sub-regions demonstrate solution type preferences in different social-ecological contexts (Fig. 7.1). We performed a chi-square test to test the association between solution type and geographical location. The results are found in Table 7.4. We found a significant association for sub-region and the four solution types with the highest difference in proportion of solution type among sub-regions (institutional, technological, knowledge, and economic) for the local-scale solution set ($n = 233$; $p = 0.035$) and for the whole solution set ($n = 396$; $p = 0.030$). No significant association was found between geographical location and all five solution types including the social, cultural, and behavioral solution type ($n = 485$; $p = 0.055$).

In the discussions below, a number of solution sub-types (for instance, ecological conservation under the technological solution type) are included in the examples and discussion. In some cases, there are a small number of each of these individual solution sub-types in our sample, while there are many examples of such activities in the sub-regions.

Institutional and governance type solutions are the most common solution type across Asia. Institutional and governance type solutions show a similar pattern at the sub-regional level and are a significant proportion of solution types regardless of geography, making up at minimum one quarter of solutions in each sub-region (Fig. 7.1). Inclusion is the highest selected solution of institutional and governance

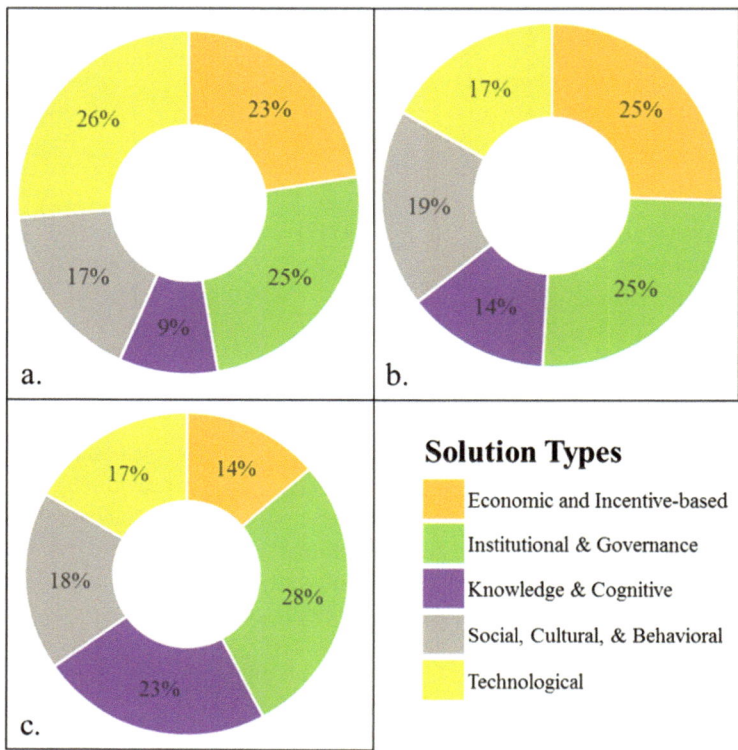

Fig. 7.1 Local-scale solutions by solution type and sub-region, in percent. (**a**) East Asia, (**b**) South Asia, (**c**) Southeast Asia

solution sub-types at the local level in all three sub-regions (Fig. 7.2). These solutions target an increase in the number of categories of actors that hold benefit rights or that are included in decision-making processes.

Social, cultural, and behavioral type solutions are represented in even proportions across the region, with an average of 18% in each sub-region. South Asia has more than double the proportion of solutions for access to social services compared to Southeast Asia. Although the result is not significant with regard to an association between sub-region and social and behavioral solution types, we can expect a higher investment in basic social services such as access to water and development activities in SEPLS in South Asia where a number of the case study countries (India, Bhutan, Nepal) have high poverty rates.

Technological solutions are preferred in East Asia at a rate of 1.5 times over other sub-regions. In East Asia, technological solutions make up 26% of the proportion of solutions selected (Fig. 7.1a), while the figures are 17% in South and Southeast Asia, respectively (Fig. 7.1b, c). The preference for technological responses in East Asia is driven by a higher emphasis on incorporating agroecological practices such as sustainable agriculture in SEPLS management, as well as renewable energy investments that are double those of Southeast Asia and are nil in South Asia.

Table 7.4 Summary of chi-square results for solution type and geographic location (East, South, and Southeast sub-regions)

Cases	Solution types tested	Number of solutions	Association with sub-regions (Pearson Chi-Square)	
			Test statistic	Asymp. Sig. (2-sided)
Local scales only Four solution types	Institutional, economic, knowledge, technological	$n = 233$	**13.543**[a]	**0.035**[b]
All scales Four solution types	Institutional, economic, knowledge, technological	$n = 396$	**13.993**[c]	**0.030**[b]
All scales Five solution types	Institutional, economic, knowledge, technological, social	$n = 485$	15.239[d]	0.055

[a]N of valid cases = 233. 0 cells (0.0%) have expected count less than 5. The minimum expected count is 8.65

[b]**Significant result**

[c]N of valid cases = 396. 0 cells (0.0%) have expected count less than 5. The minimum expected count is 21.30

[d]N of valid cases = 485. 0 cells (0.0%) have expected count less than 5. The minimum expected count is 22.09

Knowledge and cognitive type solutions are preferred in Southeast Asia at a rate of 2.5 times that of East Asia. In Southeast Asia, knowledge and cognitive type solutions make up 23% of the proportion of solutions (Fig. 7.1c), compared with just 14% in South Asia and 9% in East Asia (Fig. 7.1a, b). In Southeast Asia, local communities in SEPLS invest more often in all sub-types of knowledge and cognitive type solutions at nearly double the rate of the other sub-regions. These include capacity building (knowledge capacities), monitoring and evaluation systems (knowledge systems), and bridging knowledge forms from communities and science (knowledge integration). For instance in the Philippines, local ecological knowledge of ethnic groups is included in practical learning experiences in farmer field schools, and helps bridge local knowledge systems with new technical developments (Dang 2015). The exception is for assessments and research (knowledge gaps), which are selected in a slightly higher proportion in East Asia.

In South Asia, economic and incentive-based solutions are preferred more often than in Southeast Asia, 25% compared with 14%, respectively (Fig. 7.1b, c). While the proportion of economic-based solutions are similar in East Asia (23%) to South Asia, the distribution of solution sub-types is different among the two sub-regions. In South Asia, there is a clear preference for livelihood-based solutions that include both investments in existing livelihoods and the creation of new livelihood opportu-

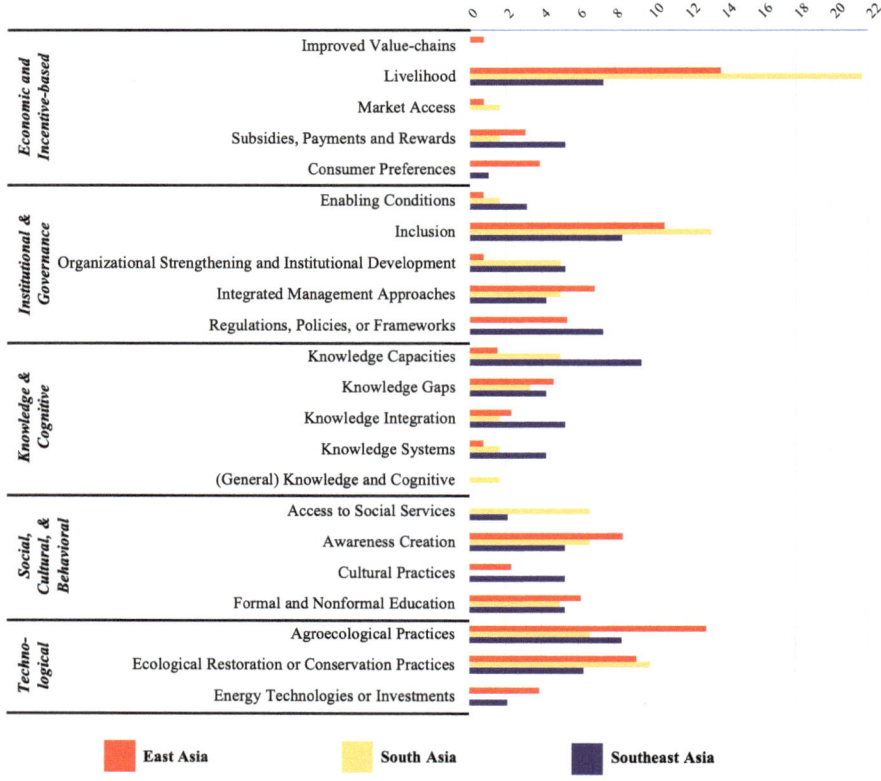

Fig. 7.2 Local-scale solutions per subcategory of solution type and per sub-region, in percent per solution type

nities. Livelihood solutions also represent the highest overall proportion of any solution sub-type.

There are a few instances of unique solutions in sub-regions. For instance, investments in improved value-chains are found only in East Asia at the local scale. One such case included the addition of a local bamboo processing industry in China (Yiping 2011). In several cases, one sub-region does not select solution types found in the other two sub-regions. Consumer preference schemes such as eco-labelling of environmentally-friendly rice in Taiwan (economic); cultural practices such as "muyong" in the Philippines that guide private forest land owners to act flexibly with regard to sustainable community resource management (social and cultural); regulations, policies, and frameworks such as community action plans at the local scale (institutional and governance); and renewable energy investments (technological) are not found in South Asia at the local scale (Fan et al. 2016; UNU-IAS 2012c). Market access interventions such as the creation of new local businesses that create markets for locally grown produce in the revitalization of small towns in

Japan are not found in Southeast Asia (Matsui et al. 2010). Access to basic social services such as clean drinking water (social and cultural) is not found in East Asia.

Taxes and user fees (economic) and financing (institutional and governance) do not appear as local-scale solutions but are found in the overall set of solutions, although such solutions are possible at the scale of local or village government.

7.3.2 Sub-regional Multi-stakeholder Coalitions for Conservation and Restoration Solutions in SEPLS

Across the IPSI network in Asia, there is strong evidence that communities are working with additional societal actors at multiple levels to achieve conservation and restoration solutions. In landscape coalitions, local communities and community-based organizations form the basis of multi-stakeholder coalitions. They represent just under one quarter of all actors and are engaged with a wide range of other stakeholders, in total 18 types of partners at different scales across 4 broad stakeholder groups (public, nongovernmental, research, community) (Fig. 7.3). In each sub-region 16 to all 18 of these partner types are found, and they include all four broad stakeholder groups in each sub-region.

However, the types of stakeholders in partnership with communities in landscape coalitions do demonstrate different sub-regional compositions depending on socio-political contexts. In Southeast Asia for instance, coalitions are more heavily made up of public sector engagement from local or national sector ministries and individual leaders at multiple scales. Such leaders are commonly an extension agent such as a District Forest Officer. Sectoral ministries are less engaged in solutions in SEPLS in South Asia in our sample.

The compositions of societal actors in coalitions in other sub-regions may reflect states of economic development. Local and national research institutions are engaged at double the rate in East Asia compared with the other two sub-regions, while foreign government engagement is not found at all, likely reflecting the more developed economies of Japan and China and more developed national research institutes and facilities. And while it's more common to find mention of individual leaders that drove conservation and restoration solutions in SEPLS outside of East Asia, local governments are engaged at double the rate in East Asia than Southeast Asia.

In South Asia, engagement of intergovernmental organizations is more prevalent, often to provide technical support or capacity building for development project implementation in SEPLS. Similarly, in South Asia engagement of the nongovernmental sector in coalitions is strong. Local and international NGOs are found at about double the rate in South and East Asia than in Southeast Asia. Civil society institutions are also a strong development partner in South Asia. In addition, community-based institutions representing indigenous and ethnic groups are found

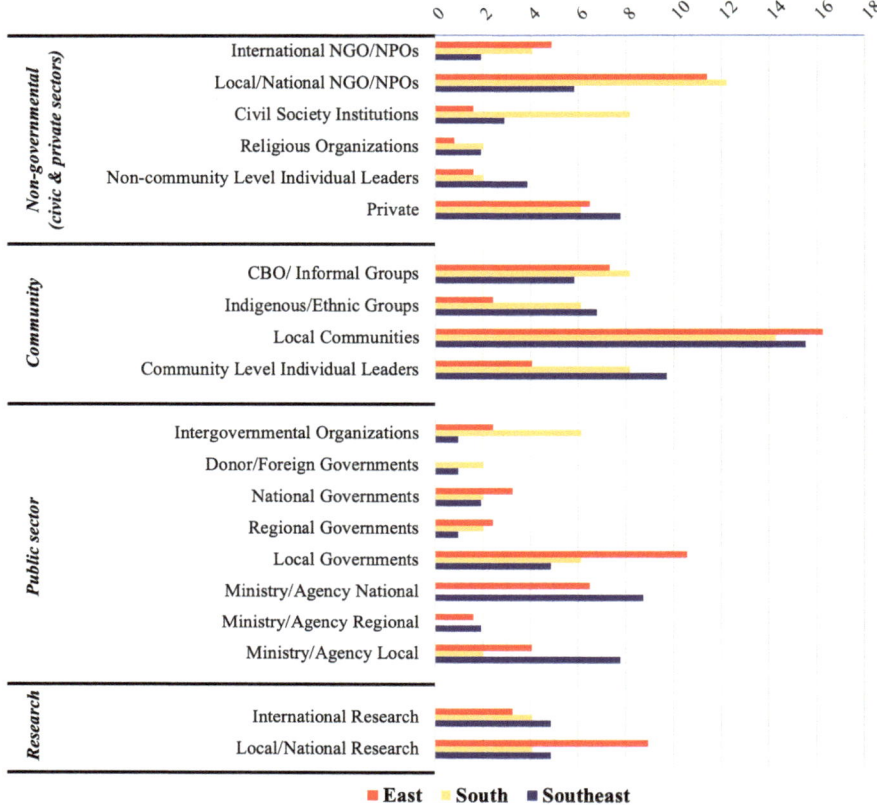

Fig. 7.3 Types of stakeholders engaged in conservation and restoration solutions in SEPLS at the local scale, by sub-region, in percent of total

at triple the rate in South and Southeast Asia than in East Asia, and participate in development and natural resource management activities in SEPLS.

Private sector and international research institutions appear to invest in cooperating or engaging in multi-stakeholder management of SEPLS at about the same rate across the three sub-regions.

7.3.3 Mosaic Landscapes of Multiple Ecosystems

The relationships of solutions across spatial scales in the eight ecosystems are shown in Fig. 7.4. Solutions tend to be planned at social and administrative scales rather than ecological scales. When planning is done at ecological scales it tends not to be at the local scales, with the exception of inland water ecosystems. For instance

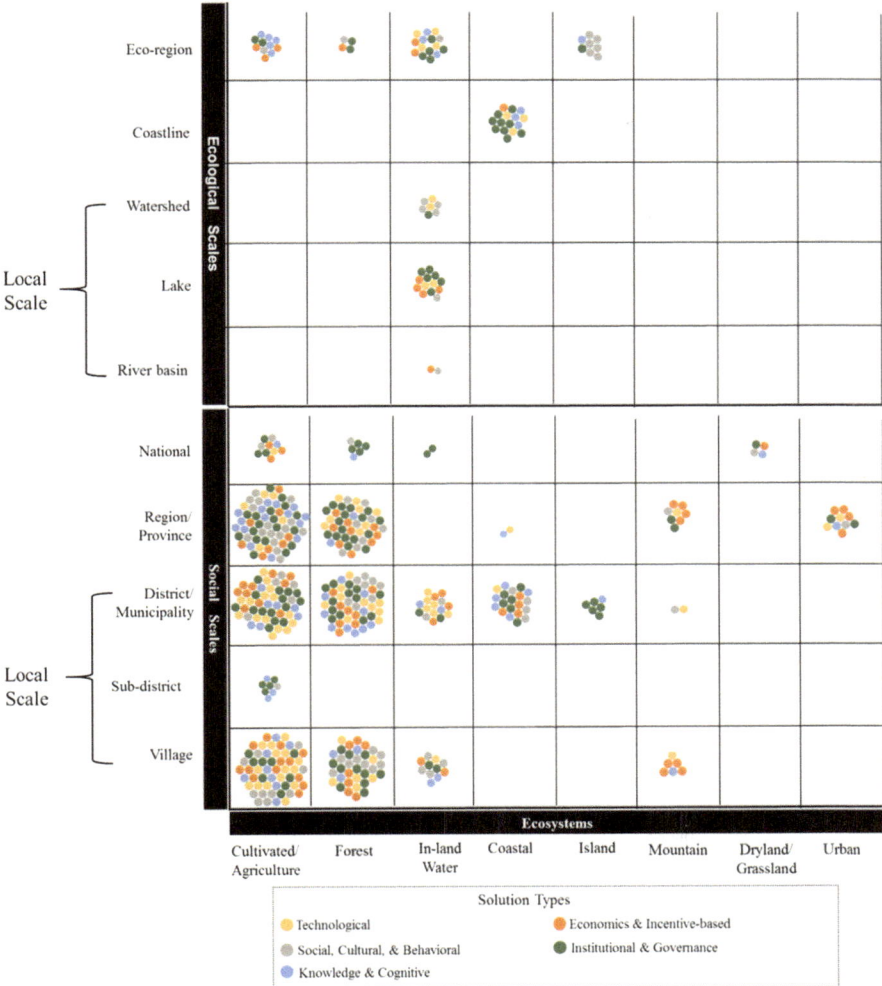

Fig. 7.4 Solutions per solution type, by social and ecological scales and ecosystem. Local social scales include village to district/municipality and ecological scales from river basin to watershed. (Source: Kozar et al. 2019 reprinted with permission from Elsevier)

in cultivated and forest ecosystems, solutions are only planned at local social and administrative scales and not at local ecological scales.

Solutions have the highest concentrations in cultivated and forest ecosystems. Although cultivated and forest ecosystems each make up roughly the same proportion of overall ecosystems found in SEPLS across the 88 cases, at the local scales, solutions are targeted at a 32% greater rate in cultivated ecosystems than forest ecosystems. In forest ecosystems, institutional and governance solution types are selected at about twice the rate of knowledge solution types. Relative to cultivated ecosystems, there is a preference for social and cultural solution types in forest ecosystems, while relative to forest ecosystems, we find a preference for technological solution types in cultivated ecosystems.

Inland water ecosystems are the only one of the eight ecosystems found in IPSI member experiences where solutions are part of sustainable use and management schemes planned at local ecological scales including river basins, freshwater lakes, and watershed or catchment areas. In inland water ecosystems, knowledge solution types are not present in implementing sustainable use and management approaches at local ecological scales, and knowledge solutions make up the lowest proportion of solution types in inland water ecosystems even when including those planned at administrative scales. Knowledge solutions are selected at two to three times less often a rate than other solution types in inland water ecosystems.

Institutional and governance solution types were not found at local scales in mountain ecosystems, while they were found in island and coastal ecosystems at a local government scale at about two times the rate to all other solution types. None of the 52 cases at local scales were found in dryland or urban ecosystems. Solutions in coastal and island ecosystems were planned only at the district or municipality level at local scales.

7.4 Discussion

7.4.1 Selecting Place-Based Solutions for Different Social-Ecological Systems

Solution types, coalitions of societal actors, and targeted ecosystems and scales do show variances per sub-region and overall patterns in the experiences of IPSI members in Asia (Figs. 7.1, 7.2, 7.3, and 7.4). We discuss some of the factors that may account for these differences in patterns by sub-region.

The preference for technological responses in East Asia, particularly for agroecological solutions (Fig. 7.2), can be ascribed to the recognition of the impacts of modernized agriculture in countries such as Japan, Korea, Taiwan, and China. Japan and China have among some of highest rates of pesticide use worldwide,[3] and the impacts of these practices on key species important for high-value crops triggered local communities to shift to more eco-friendly farming practices. In Taiwan, farmers developed new production systems based on ecological practices (Chao 2018). In Japan and China, species such as the Oriental White Stork in the Hyogo Tajima region of Japan and the Crested Ibis in China inspired the adoption of agroecological farming practices that would restore and create a habitat for these species to

---

[3] Japan, Taiwan, and China are in the highest bracket of countries worldwide, and Korea follows in the second highest country bracket, per tonnes of active ingredients of pesticides, averaged 1990–2016. Both China and Japan are in the top ten countries. http://www.fao.org/faostat/en/#data/RP/visualize (accessed January 29, 2019).

cohabitate in cultivated ecosystems (Ichikawa 2012; Ohsako 2011). There is also a relatively higher engagement of national and local research institutes in East Asia that may support the research needed to develop localized technical solutions.

The higher selection of solutions such as learning processes or training programs (Figs. 7.1 and 7.2) demonstrates the preference for more knowledge-based approaches to achieving agroecological farming and forest harvesting in Southeast Asia, whereas we saw a more technical approach in East Asia to meet the same aim. Knowledge investments in Southeast Asia include: the practical learning and experimental spaces provided by farmers' field schools in Vietnam (Dang 2015); participatory learning through field schools and experimentation (Setiawan and Khumairoh 2014), and action research, participatory monitoring and learning groups for agroforestry in Indonesia (Amaruzaman et al. 2018); formation and strengthening of community management groups and their capacities to advocate for policy change in tandem with promotion of ecological farming and exchange of traditional seed varieties in the Philippines (MRDC and Tebtebba Foundation 2012); and participatory monitoring by community members in Cambodia (Costello and Vorsak 2011). In Southeast Asia, there are more government sectoral ministries engaged in knowledge-based solutions for sustainable use and management. This may help explain why we find a high number of government extension and learning programs for sustainable resource management at local scales.

The preference for livelihood-based solutions that also address food and nutrition security as the largest proportion of economic and incentive-based solution types in South Asia (Figs. 7.1 and 7.2) is illustrated by: investments in new cash crops for income and food sufficiency in Nepal; investments in cooperative farming of fish and single cell algal protein, and proposals for integration of horticulture practices with indigenous fruit species in India; and proposals to diversify livelihood options to reduce poverty in Bhutan (Pandit et al. 2016; ICIMOD 2017; Tekale et al. 2012). The high preference for investments in livelihoods, access to social services, and organizational strengthening may reflect the more basic development needs of the region and the higher incidence of poverty. This focus on items such as access to water and capacity building of local institutions builds the foundation for the sustainable management of SEPLS. There is also a higher abundance of local NGOs and CBOs that may support small development activities to invest in livelihoods and social services as well as civil society institutions that may be able to advocate for citizen needs. Box 7.1 illustrates a profile of multi-stakeholder governance of SEPLS in Nepal, typical of findings in South Asia, and discusses how the composition of societal-based coalitions and solution types interact at local scales.

Box 7.1 Community Engagement in Navigating Solutions to the Loss of BES in SEPLS in Nepal

Nepal is a biodiversity-rich country with a multicultural population representing significant ethnic diversity. Nature is considered a complementary part of everyday life, and the majority of people's livelihoods depend on natural resources in a rural economy based on subsistence agriculture. Communities are dependent on the forest for fuelwood, fodder, and timber, and inland freshwater resources for fishing, irrigation, and daily water consumption for drinking, washing, and bathing. SEPLS in Nepal face social and ecological challenges such as low education levels, high use of chemical fertilizer, overharvesting, insufficient infrastructure and social services, and an increasing population density that has led to rising food insecurity challenges.

Nonetheless, local communities are contributing to the effective management of local natural resources through their traditional knowledge and a variety of localized resource-based management systems such as the Federation of Community Forestry User groups, farmer to farmer and community-based cooperatives, community drinking water management groups, and community-based protected area management (Adhikari 2011; ILEC 2012; UNDP 2014b; Pandit et al. 2016). They often do this through livelihood (economic) and inclusion (institutional) solution sub-types, which are found with the highest frequency in South Asia. Moreover, marginal and indigenous communities undertake entrepreneurial activities through sustainable use of the resources found in the surroundings in which they inhabit. For instance, their beliefs in maintaining an eternal relationship with nature are evident in their development of sustainable solutions for improved local community livelihoods through strategies such as domestication of wild medicinal aromatic plant species (Pandit et al. 2016).

Similar to South Asia as a whole, in Nepal there is an abundance of local NGOs, civil society organizations such as Ward Citizen Forums, and community-based organizations including youth clubs, which are engaged in conservation and restoration activities. These organizations are also actively involved in realizing the significant role of the community in raising awareness to address issues of biodiversity conservation and sustainable use of resources through trainings, workshops, exposure visits, posters and pamphlets, as well as for mobilization and strengthening of community groups (Adhikari 2011; ILEC 2012; UNDP 2014b; Pandit et al. 2016). In Nepal, as in South Asia, there is a strong presence of intergovernmental agencies engaged in project management, and that provide technical support for implementation.

Technical solutions such as ecological conservation and restoration are found most frequently in South Asia. In Nepal, forest area covers 40% of the total land area, and forest resources make a major contribution to addressing poverty and enhancing the resilience of local communities. Nepal has become

(continued)

> **Box 7.1** (continued)
>
> a successful model of the community forestry management system, which is well known for promoting sustainable management and restoration of the forest. Improving forest-based livelihoods depends on community-based forest management. The system relies on community-based governance through beneficial linkages with concerned stakeholders such as the District Forest Office (DFO), the Federation of Community Forestry User Groups Nepal (FECOFUN), the Department of Forestry (DOF), national and village development committees, and community groups of women, youth and cooperatives. These networks of stakeholders actively work together to provide capacity building to local communities to implement the management schemes (Adhikari 2011). Traditional knowledge is encouraged in meetings, planning, and decision-making processes from the community to the district level forum.

Some specific solutions can be explained by the choices of international research institutions, international NGOs, and private sector actors to invest in programs that aim to meet global conservation and restoration aims and secure international value-chains by increasing benefit streams to local residents of SEPLS. The subsidies, payments, and rewards solutions in Southeast Asia may reflect such international investments in the intensive piloting and promotion of payment for ecosystems services (PES) schemes in countries such as Indonesia, Cambodia, and the Philippines (see for instance Amaruzaman et al. 2018; Costello and Vorsak 2011).

7.4.2   Meeting the New Challenges for Sustainable Use and Management of SEPLS

IPSI's experiences in revitalizing sustainable use and management of SEPLS do demonstrate strong efforts across Asia to find solutions through inclusion, which have in some cases strengthened multi-level governance of SEPLS. Place-based solutions emphasize institutional and governance solution types as the most selected solution type, and especially solutions for inclusion, that aim to bring in more collaborative decision-making modes or more equitable benefit sharing of ecosystem services. For instance, inclusion of local people in defining rights to land use can lead to strengthening regulatory systems and institutions governing customary law, connecting local institutions with national forest policy (Shohibuddin and Aoyama 2009).

Moreover, because place-based solutions are already based in collaborative decision-making processes by coalitions of societal actors, these solutions add value to best available knowledge of conservation and restoration solutions and what works across multiple scales. Indeed, existing private sector engagement and engagement of international and local research institutions, while present across the region, can be enhanced by stakeholder coalitions that reach out to engage additional actors and continue to encourage inclusive approaches.

Research on governance modes to address the interlinked drivers of a loss of BES increasingly calls for governance by "imagination," or to anticipate what will be needed in an uncertain future shaped by as yet undetermined impacts of climate change and biodiversity loss (Burch et al. 2019). These new states of sustainable use and management need to be resilient states that anticipate and imagine ecosystem service needs in the future as well as the present and that account for potential climate change impacts. In a recent review, Chiu et al. (2018) found that future resilience strategies would get the best return from those solutions that demonstrate co-benefits, or in the case of SEPLS, those that meet the revitalization, adaptation and innovative production and livelihood needs of local residents, while conserving and restoring landscapes for climate mitigation and biodiversity conservation benefits.

Societal actors demonstrate preferences for solution types to reversing the loss of BES in SEPLS while embracing all solution types across ecosystems. In current management practices in mosaic SEPLS, stakeholders rely on different tool-kits for managing BES on a local scale to achieve the same international targets such as the Aichi Biodiversity Targets or Sustainable Development Goals depending on the social-ecological system, stakeholder coalitions, and social-cultural context. This suggests that sustainable use and management to conserve and restore BES on multiple scales can be achieved through varying management strategies.

Experiences in the IPSI network also demonstrate that investments in a spectrum of solution types including cognitive, economic, social and cultural, and technological are needed to meet the challenges of new and effective sustainable use and management approaches. The type of conservation and restoration solutions appropriate in different social-ecological contexts may depend on the composition of stakeholder coalitions, planning systems at social and ecological scales, and the mosaic ecosystem character of SEPLS.

We found overall that institutional and governance solution types are the most frequent at local scales and at all scales combined. This is a good start to developing the coordinated responses at multiple scales that can consider the continuing adaptation and management of ecosystem services in the future (MA 2005), and this can be achieved with the harmonization of solutions in institutions and governance together with solutions in all solution types. Mixes of different solution types with the intention to achieve synergies among biodiversity and production benefits can help address the feedbacks across scales, and among land uses and users, in complex social-ecological systems such as SEPLS.

Further, many of the technological solutions in IPSI member experiences have the potential for multiple benefits, such as renewable energy technologies or ecological farming practices. If these are intentionally combined for better environmental decision-making with knowledge-based solutions, and with institutional and social solutions that can provide supportive policy, management strategies and education and awareness, this will help ensure the realization of these co-benefits in future sustainable use and management of SEPLS.

7.4.3   Place-Based Solutions for Sustainable Use and Management of Production Landscapes and Seascapes

Understanding which place-based solution types are preferred in different social-ecological contexts helps improve more purposeful design of SEPLS for multiple benefits, particularly in considering where optimal ecosystem services are not present and where different solution types might increase synergies or reduce trade-offs among land-use choices. Social-cultural approaches are more popular in sustainable use and management of common resources such as forest, while more technical approaches are preferred in management of cultivated lands, which may be more tied to individual farmers' decisions. New sustainable use and management approaches can utilize existing stakeholder preferences exemplified in place-based solutions, but also examine how widening values of multiple stakeholders and coordinated responses might create benefits from inclusion of a different solution type or efforts at a different scale.

For instance, greater engagement of national research institutes and local and national NGOs in South Asia might help incorporate more tailored technical solutions where appropriate. While catchment management and integrated lake basin management approaches have benefited from investments in multi-stakeholder governance, knowledge or cognitive solutions that are currently less utilized in the East and South Asia sub-regions could lift inadequate use of knowledge. Linking integrated management (an institutional solution) with monitoring and evaluation systems (knowledge solution) could help bring about more robust SEPLS management that integrates multiple knowledge forms from indigenous and practical sources, and creates local information systems such as participatory monitoring that can monitor feedbacks across scales.

Some limitations of sustainable use and management approaches in SEPLS can be improved. While integrated landscape and transboundary approaches are gaining popularity, we found that place-based solutions are targeted mainly in cultivated and forest ecosystems. Further, there hasn't yet been a shift to planning the majority of agricultural and forest ecosystem transformations at landscape or ecological scales because the institutional architecture is embodied in village and district or municipality governance systems, while ecosystem services may not have any bearing on these scales. This may be true particularly in Southeast Asia where the most common composition of stakeholder coalitions in SEPLS is the interaction of sectoral ministries with communities through planning at social and administrative scales. Further, mountain ecosystems, drylands, and urban areas do not receive any of the benefits of planning at an eco-regional scale that might consider the flows of ecosystem services. In some cases, dryland, urban, and island ecosystems may benefit from more integrated landscape approaches at administrative scales as they are typically tackled at higher scales from district to national levels. Island ecosystems may correspond to a district or municipality and benefit from synching of ecological and social scales. This would entail more deliberate spatial planning that bridges social and ecological knowledge, and that gives consideration to the interactions of drivers and solution types at different scales and in different ecosystems from among the full suite of technical, social, institutional, economic, and knowledge-based solutions.

Deliberate place-based planning for sustainable use and management that will enable future resilient SEPLS should take advantage of community-led innovations, multi-level governance approaches, and solution mixes that include social, cultural, and knowledge-based approaches along with context-dependent technical and economic solutions (Bohnet and Beilin 2015). As urban areas expand, planning for these areas that incorporates multiple solution types can enhance their resilience, especially when solution mixes are selected for multiple goals, and for future resilience across urban-rural landscapes (Wendling et al. 2018; Eggermont et al. 2015). This kind of design fits with broader approaches to planning solutions in SEPLS in ways that support policy design for the Anthropocene, where synergies among solutions are key to balancing trade-offs and staying within planetary boundaries through connected places (Sterner et al. 2019).

7.5   Conclusions

The drivers of overconsumption, urbanization, modernization of agriculture, underutilization and industrialization, among others, and their interactions and cumulative effects are accelerating the loss of BES in SEPLS in Asia, and causing imbalances in formerly harmonious approaches to sustainable use and management. Externally influenced drivers and societal actors from all scales are part of the challenges but are also part of the solutions. IPSI members demonstrate that communities in SEPLS across Asia are working with multiple partners at different scales to revitalize social-ecological production landscapes and seascapes in ways that will address the drivers of BES loss. In many cases, they are also helping to revitalize a state of sustainable use and management that delivers ecosystem service benefits in the form of production income, dietary diversity, and increased health and well-being, while providing conservation and restoration benefits that benefit ecosystems and environmental health at multiple scales.

Through working with multiple stakeholders to devise institutional, cognitive, economic, social and cultural, and technological responses across ecosystems and scales, communities in SEPLS in Asia are responding to the need to devise solutions across ecosystems and scales and to consider feedbacks among social-ecological systems to revitalize a balance among the multiple ecosystems services for food, forests, biodiversity, and livelihoods. As demonstrated by the wide breadth of societal actors working jointly on solutions in SEPLS, the representation across stakeholder groups, and varying composition of coalitions, the IPSI network is contributing to the best available knowledge that reflects the pluralistic values of multiple stakeholders in identifying which solutions are preferable in different social-ecological systems.

Understanding which solution types societal actors have chosen to implement or have proposed in SEPLS in different sub-regions and social-ecological contexts in Asia gives us an understanding as to what might be the criteria for effectiveness applied by societal actors. Sharing the knowledge of which solution types are preferred in different social-ecological contexts may help societal actors make better informed decisions in weighing the appropriateness of different solutions in SEPLS. The framework proposed here for analyzing solutions by type, scale, and ecosystem can be done in a local context through a transdisciplinary approach to help coalitions of actors take note of gaps and devise place-based sustainable use and management approaches for conservation and restoration of BES in SEPLS.

Our results and findings are limited by the small number of cases available from certain ecosystems such as drylands and the relative lack of cases in Satoumi (coastal ecosystems). We were also limited by a lack of spatial data that could inform the placement of solutions in mosaic ecosystems of SEPLS. Future studies may seek to measure the performance and impact of different solutions and their combinations, and to disaggregate the solutions by existing and proposed in further analysis.

Acknowledgements We would like to thank Geetha Mohan at UNU-IAS and two anonymous reviewers for their comments on an earlier version of the manuscript. We thank the International Satoyama Initiative and the Ministry of Environment Japan for their support of this work.

References

Aadrean UN (2017) Small-clawed otters (Aonyx cinereus) in Indonesian rice fields: latrine site characteristics and visitation frequency. Ecol Res 32(6):899–908

Adhikari S (2011) Community forestry in Nepal. Case studies, International Partnership for the Satoyama Initiative. https://satoyama-initiative.org/casestudies

Akça E, Takashi K, Sato T (2015) Development and success, for whom and where: the Central Anatolian case. Land restoration: reclaiming landscapes for a sustainable future. Academic Press, Oxford, pp 533–541

Amaruzaman S, Lusiana B, Leimona B, Tanika L, Hendrawan DC (2018) Strengthening small-holder resilience and improving ecosystem services provision in Indonesia: experience from Buol District, Central Sulawesi. Case studies, International Partnership for the Satoyama Initiative. https://satoyama-initiative.org/casestudies

Bélair C, Ichikawa K, Wong BYL, Mulongoy KJ (2010) Sustainable use of biological diversity in socio-ecological production landscapes. Background to the Satoyama Initiative for the benefit of biodiversity and human well-being. Secretariat of the Convention on Biological Diversity, Montreal. Technical Series no. 52, 184 pages

Berglund BE, Kitagawa J, Lageras P, Nakamura K, Sasaki N, Yasuda Y (2014) Traditional farming landscapes for sustainable living in Scandinavia and Japan: global revival through the Satoyama Initiative. AMBIO 43:559–578

Bohnet IC, Beilin R (2015) Editorial: pathways towards sustainable landscapes. Sustain Sci 10:187–194

Burch S, Gupta A, Inoue CYA, Kalfagianni A, Persson A, Gerlak AK, Ishii A, Patterson J, Pickering J, Scobie M, Van der Heijden J, Vervoort J, Adler C, Bloomfield M, Djalante R, Dryzek J, Galaz V, Gordon C, Harmon R, Jinnah S, Kim RE, Olsson L, Van Leeuwen J, Ramasar V, Wapner

P, Zondervan R (2019) New directions in earth system governance research. Earth Syst Gov 1:100006

Chao JT (2018) Converting pests into allies in tea farming—a SEPL case in Hualien. Case studies, International Partnership for the Satoyama Initiative. https://satoyama-initiative.org/casestudies

Chiu B, Zusman E, Lee S, Jian H (2018) The co-benefits of integrated solutions in Asia: an analysis of governance challenges and enablers. In: Zusman E, Amanuma N (eds) Governance for integrated solutions to sustainable development and climate change: from linking issues to aligning interests. Institute for Global Environmental Strategies (IGES), Tokyo, pp 21–37

Cockburn J, Cundill G, Shackleton S, Rouget M (2018) Towards place-based research to support social–ecological stewardship. Sustainability 10:1434

Costello L, Vorsak B (2011) Natural resource management in the critical habitat of Western Siem Pang. Case studies, International Partnership for the Satoyama Initiative. https://satoyama-initiative.org/casestudies

Dang KT (2015) The importance of the Farmers Field School approach: a case study—Farmers Field School practical training programs, Vietnam. Case studies, International Partnership for the Satoyama Initiative. https://satoyama-initiative.org/casestudies

Dicks LV, Wright HL, Ashpole JE, Hutchison J, McCormack CG, Livoreil B, Zulka KP, Sutherland WJ (2016) What works in conservation? using expert assessment of summarised evidence to identify practices that enhance natural pest control in agriculture. Biodivers Conserv 25:1383–1399

Duraiappah AK, Tanyi AST, Brondizio ES, Kosoy N, O'Farrell PJ, Prieur-Richard AH, Subramanian SM, Takeuchi KC (2014) Managing the mismatches to provide ecosystem services human well-being: a conceptual framework for understanding the new commons. Curr Opin Environ Sustain 7:94–100

Eggermont H, Balian E, Azevedo JMN, Beumer V, Brodin T, Claudet J, Fady B, Grube M, Keune H, Lamarque P, Reuter K, Smith M, van Ham C, Weisser WW, Le Roux X (2015) Nature-based solutions: new influence for environmental management and research in Europe. GAIA 24(4):243–248

Fan ML, Yu CY, Lin L (2016) Indicator species for agrobiodiversity in rice paddy fields: research and its application to a new eco labelling scheme in eastern rural Taiwan. Case studies, International Partnership for the Satoyama Initiative. https://satoyama-initiative.org/casestudies

Freeman OE, Duguma LA, Minang PA (2015) Operationalizing the integrated landscape approach in practice. Ecol Soc 20(1):24

Gu H, Subramanian SM (2014) Drivers of change in socio-ecological production landscapes: implications for better management. Ecol Soc 19(1):41

Hashimoto S, Nakamura S, Saito O, Kohsaka R, Kamiyama C, Tomiyoshi M, Kishioka T (2015) Mapping and characterizing ecosystem services of social–ecological production landscapes: case study of Noto, Japan. Sustain Sci 10(2):257–273

Havas J, Saito O, Hanaki K, Tanaka T (2016) Perceived landscape values in the Ogasawara Islands. Ecosyst Serv 18:130–140

Hernández-Morcillo M, Burgess P, Mirck J, Pantera A, Plieninger T (2018) Scanning agroforestry-based solutions for climate change mitigation and adaptation in Europe. Environ Sci Policy 80:44–52

Ichikawa K (2012) China: rural communities in cohabitation with the Crested Ibis in Yang County, Shaanxi Province. Case studies, International Partnership for the Satoyama Initiative. https://satoyama-initiative.org/casestudies

ICIMOD (International Centre for Integrated Mountain Development), Nepal, Royal Society for Protection of Nature (2017) A multi-dimensional assessment of ecosystems and ecosystem services in Barshong, Bhutan. Case studies, International Partnership for the Satoyama Initiative. https://satoyama-initiative.org/casestudies

ILEC (International Lake Environment Committee Foundation) (2012) Governance at community level for conserving Himalayan Lake Rupa in Kaski District of Nepal. Case studies, International Partnership for the Satoyama Initiative. https://satoyama-initiative.org/casestudies

IPBES (Intergovernmental Science-Policy Platform on Biodiversity and Ecosystem Services) (2015) Preliminary guide regarding diverse conceptualization of multiple values of nature and its benefits, including biodiversity and ecosystem functions and services (deliverable 3(d)). IPBES/4/INF/13

IPBES (Intergovernmental Science-Policy Platform on Biodiversity and Ecosystem Services) (2018) Summary for policymakers of the regional assessment report on biodiversity and ecosystem services for Asia and the Pacific of the Intergovernmental Science-Policy Platform on Biodiversity and Ecosystem Services. IPBES Secretariat, Bonn. (41 pages)

Jacobs S, Dendoncker N, Martín-López B, Barton DN, Gomez-Baggethun E, Boeraeve F, McGrath FL, Vierikko K, Geneletti D, Sevecke KJ, Pipart N, Primmer E, Mederly P, Schmidt S, Aragão A, Baral H, Bark RH, Briceno T, Brogna D, Cabral P, De Vreese R, Liquete C, Mueller H, Peh KSH, Phelan A, Rincón Ruiz A, Rogers SH, Turkelboom F, Van Reeth W, van Zanten BT, Wam HK, Washbourne CL (2016) A new valuation school: integrating diverse values of nature in resource and land use decisions. Ecosyst Serv 22(b):213–220

Katayama N, Baba YG, Kusumoto Y, Tanaka K (2015) A review of postwar changes in rice farming and biodiversity in Japan. Agric Syst 132:73–84

Khan K, Ikram, M, Takahi, S, Iqbal R (2017) Socio-ecological landscape change as a preamble to mountainous urban watershed rejuvenation, Kanshi of the Jhelum River Basin, Potohar Plateau, Pakistan. Case studies, International Partnership for the Satoyama Initiative. https://satoyama-initiative.org/casestudies

Kieninger PR, Yamaji E, Penker M (2011) Urban people as paddy farmers: the Japanese Tanada Ownership System discussed from a European perspective. Renew Agric Food Syst 26(4):328–341

Knight C (2010) The discourse of "encultured nature" in Japan: the concept of Satoyama and its role in 21st century nature conservation. Asian Stud Rev 34(4):421–441

Kohsaka R, Shih W, Saito O, Sadohara S (2013) Local assessment of Tokyo: Satoyama and satoumi—traditional landscapes and management practices in a contemporary urban environment. In: Elmqvist T, Fragkias M, Goodness J, Güneralp B, Marcotullio PJ, McDonald RI, Parnell S, Schewenius M, Sendstad M, Seto KC, Wilkinson C (eds) Urbanization, biodiversity and ecosystem services: challenges and opportunities: a global assessment. Springer, Dordrecht, pp 93–105

Kozar R, Galang E, Alip A, Sedhain J, Subramanian SM, Saito O (2019) Multi-level networks for sustainability solutions: the case of the International Partnership for the Satoyama Initiative. Curr Opin Environ Sustain 39:123–134

Kumar BM, Takeuchi K (2009) Agroforestry in the Western Ghats of peninsular India and the satoyama landscapes of Japan: a comparison of two sustainable land use systems. Sustain Sci 4(2):215–232

Li S, Li X (2016) Progress and prospect on farmland abandonment. Dili Xuebao/Acta Geogr Sin 71(3):370–389

MA (Millennium Ecosystem Assessment) (2005) Millennium Ecosystem Assessment. Ecosystems and human well-being: synthesis. Island Press, Washington, DC. (155 pages)

Marady I, Sokla H, Mary S (2011) Role and involvement of the commune council in community forestry activities in Domnak Neak Tathmor Puan. Case studies, International Partnership for the Satoyama Initiative. https://satoyama-initiative.org/casestudies

Matsui T, Kawashima T, Kasahara T (2010) Town revitalization making the most of natural landscape and traditions of Kanakura Wajima City, Ishikawa Prefecture, Japan. Case studies, International Partnership for the Satoyama Initiative. https://satoyama-initiative.org/casestudies

Morimoto Y (2011) What is Satoyama? Points for discussion on its future direction. Landsc Ecol Eng 7(2):163–171

MRDC (Montañosa Research and Development Center), Tebtebba Foundation (2012) Role of traditional knowledge in strengthening socio-ecological production landscapes. Case studies, International Partnership for the Satoyama Initiative. https://satoyama-initiative.org/casestudies

Ohsako Y (2011) Reintroduction project of the Oriental White Stork for coexistence with humans in satoyama areas, Hyogo, Japan. Case studies, International Partnership for the Satoyama Initiative. https://satoyama-initiative.org/casestudies

Okayasu S, Matsumoto I (2013) Contributions of the Satoyama initiative to mainstreaming sustainable use of biodiversity in production landscapes and seascapes. Institute for Global Environmental Strategies (IGES), Tokyo. (44 pages)

Oliver TH, Heard MS, Isaac NJB, Roy DB, Procter D, Eigenbrod F, Freckleton R, Hector A, Orme CDL, Petchey OL, Proença V, Raffaelli D, Suttle KB, Mace GM, Martín-López B, Woodcock BA, Bullock JM (2015) Biodiversity and resilience of ecosystem functions. Trends Ecol Evol 30(11):673–684

Osawa T, Kohyama K, Mitsuhashi H (2016) Multiple factors drive regional agricultural abandonment. Sci Total Environ 542:478–483

Pandit BH, Lopez-Casero MF, Pandit NR, Aryal NK (2016). Strengthening local capacity for conserving medicinal plants and improving livelihoods through domestication and integration of LBSAP in planning process. In: UNU-IAS and IGES (ed) Mainstreaming concepts and approaches of socio-ecological production landscapes and seascapes into policy and decision-making (Satoyama initiative thematic review, vol 2). United Nations University Institute for the Advanced Study of Sustainability, Tokyo, pp 85–95

Pascual U, Balvanera P, Díaz S, Pataki G, Roth E, Stenseke M, Watson RT, Başak Dessane E, Islar M, Kelemen E, Maris V, Quaas M, Subramanian SM, Wittmer H, Adlan A, Ahn S, Al-Hafedh YS, Amankwah E, Asah ST, Berry P, Bilgin A, Breslow SJ, Bullock C, Cáceres D, Daly-Hassen H, Figueroa E, Golden CD, Gómez-Baggethun E, González-Jiménez D, Houdet J, Keune H, Kumar R, Ma K, May PH, Mead A, O'Farrell P, Pandit R, Pengue W, Pichis-Madruga R, Popa F, Preston S, Pacheco-Balanza D, Saarikoski H, Strassburg BB, van den Belt M, Verma M, Wickson F, Yagi N (2017) Valuing nature's contributions to people: the IPBES approach. Curr Opin Environ Sustain 26–27:7–16

Plieninger T, van der Horst D, Schleyer C, Bieling C (2014) Sustaining ecosystem services in cultural landscapes. Ecol Soc 19(2):59

Plieninger T, Kohsaka R, Bieling C, Hashimoto S, Kamiyama C, Kizos T, Penker M, Kieninger P, Shaw BJ, Sioen GB, Yoshida Y, Saito O (2018) Fostering biocultural diversity in landscapes through place-based food networks: a "solution scan" of European and Japanese models. Sustain Sci 13:219–233

Putra RE, Nakamura K (2009) Foraging ecology of a local wild bee community in an abandoned Satoyama system in Kanazawa, Central Japan. Entomol Res 39(2):99–106

Queiroz C, Beilin R, Folke C, Lindborg R (2014) Farmland abandonment: threat or opportunity for biodiversity conservation? A global review. Front Ecol Environ 12(5):288–296

Sakurai R, Ota T, Uehara T, Nakagami K (2016) Factors affecting residents' behavioral intentions for coastal conservation: case study at Shizugawa Bay, Miyagi, Japan. Mar Policy 67:1–9

Setiawan A, Khumairoh U (2014) Diversifying forage composition to improve milk production and quality through participatory learning. Case studies, International Partnership for the Satoyama Initiative. https://satoyama-initiative.org/casestudies

Shimada D (2015) Multilevel natural resources governance based on local community: a case study of seminatural grassland in Tarōji, Nara, Japan. Int J Commons 9(2):486–509

Shohibuddin M, Aoyama G (2009) Creation and management of diverse secondary forest in Central Sulawesi, Indonesia. Case studies, International Partnership for the Satoyama Initiative. https://satoyama-initiative.org/casestudies

Springmann M, Mason-D'croz D, Robinson S, Garnett T, Charles H, Godfray J, Scarborough P (2016) Global and regional health effects of future food production under climate change: a modelling study. Lancet 387(10031):1937–1946

Sterner T, Barbier EB, Bateman I, van den Bijgaart I, Crépin AS, Edenhofer O, Fischer C, Habla W, Hassler J, Johansson-Stenman O, Lange A, Polasky S, Rockström J, Smith HG, Steffen W, Wagner G, Wilen JE, Alpízar F, Azar C, Carless D, Chávez C, Coria J, Engström G, Jagers SC, Köhlin G, Löfgren A, Pleijel H, Robinson A (2019) Policy design for the Anthropocene. Nat Sust 2:14–21

Subramanian SM, Kaoru I, Ayako K (2015) Enhancing knowledge for better management of socio-ecological production landscapes and seascapes: appropriate tools and approaches for effective action. In: UNU-IAS and IGES (ed) Enhancing knowledge for better management of socio-ecological production landscapes and seascapes (SEPLS) (Satoyama initiative thematic review, vol 1). United Nations University Institute for the Advanced Study of Sustainability, Tokyo, pp 1–7

Subramanian SM., Chakraborty S, Ichikawa K (2016) Toward mainstreaming concepts and approaches of socio-ecological production landscapes and seascapes (SEPLS): lessons from the field. In: UNU-IAS and IGES (ed) Mainstreaming concepts and approaches of socio-ecological production landscapes and seascapes into policy and decision-making (Satoyama initiative thematic review, vol 2). United Nations University Institute for the Advanced Study of Sustainability, Tokyo. pp 1–12

Subramanian SM, Chakraborty S, Leimona B (2017) Sustainable livelihood options in SEPLS for human well-being. In: UNU-IAS and IGES (ed) Sustainable livelihoods in socio-ecological production landscapes and seascapes (Satoyama initiative thematic review, vol 3). United Nations University Institute for the Advanced Study of Sustainability, Tokyo, pp 1–11

Subramanian SM, Yiu E, Leimona B (2018) Enhancing effective area-based conservation through the sustainable use of biodiversity in socio-ecological production landscapes and seascapes (SEPLS). In: UNU-IAS and IGES (ed) Sustainable use of biodiversity in socio-ecological production landscapes and seascapes and its contribution to effective area-based conservation (Satoyama initiative thematic review, vol 4). United Nations University Institute for the Advanced Study of Sustainability, Tokyo, pp 1–13

Sutherland WJ, Gardner T, Bogich TL, Bradbury RB, Clothier B, Jonsson M, Kapos V, Lane SN, Möller I, Schroeder M, Spalding M, Spencer T, White PCL, Dicks LV (2014) Solution scanning as a key policy tool: identifying management interventions to help maintain and enhance regulating ecosystem services. Ecol Soc 19(2):3

Sutherland WJ, Dicks LV, Ockendon N, Smith RK (2017) What works in conservation 2017. Open Book Publishers, Cambridge. (445 pages)

Takeuchi K, Ichikawa K, Elmqvist T (2016) Satoyama landscape as social-ecological system: historical changes and future perspective. Curr Opin Environ Sustain 19:303

Tekale NS, Kodarkar MS, Karnik P, Singh R (2012) Integrated lake basin management (ILBM), impacts on biodiversity and child malnutrition: a case study of tribal belt in western part of India. Case studies, International Partnership for the Satoyama Initiative. https://satoyama-initiative.org/casestudies

Tomita A, Nakura Y, Ishikawa T (2015) Review of coastal management policy in Japan. J Coast Conserv 19(4):393–404

Toyooka City (2012) Community development to live in harmony with the Oriental White Stork in Toyooka City, Hyogo, Japan. Case studies, International Partnership for the Satoyama Initiative. https://satoyama-initiative.org/casestudies

Tsuchiya K, Okuro T, Takeuchi K (2013) The combined effects of conservation policy and comanagement alter the understory vegetation of urban woodlands: a case study in the Tama Hills area, Japan. Landscape Urban Plan 110(1):87–98

UNDP (United Nations Development Programme) (2014a) Communities in action for landscape resilience and sustainability—the COMDEKS Programme. United Nations Development Programme, New York

UNDP (United Nations Development Programme) (2014b) COMDEKS project: Makawanpur District. Communities in action for landscape resilience and sustainability—the COMDEKS Programme. United Nations Development Programme, New York, pp 110–121

UNDP (United Nations Development Programme) (2016) A community-based approach to resilient and sustainable landscapes: lessons from phase II of the COMDEKS Programme. UNDP, New York

UNU-IAS (United Nations University Institute for the Advanced Study of Sustainability) (2012a) South Korea: traditional rural landscape "Maeul". In: Ichikawa K (ed) Socio-ecological production landscapes in Asia. United Nations University Institute for the Advanced Study of Sustainability, Tokyo, pp 29–33

UNU-IAS (United Nations University Institute for the Advanced Study of Sustainability) (2012b) Socio-ecological production landscapes in Asia. In: Ichikawa K (ed). United Nations University Institute for the Advanced Study of Sustainability, Tokyo

UNU-IAS (United Nations University Institute for the Advanced Study of Sustainability) (2012c) Philippines: a combination of rice terraces, Swidden and Muyong (Privately Owned Forests) in the Province of Ifugao UNU-IAS. 2012. In: Ichikawa K (ed) Socio-ecological production landscapes in Asia. United Nations University Institute for the Advanced Study of Sustainability, Tokyo, pp 75–78

UNU-IAS (United Nations University Institute for the Advanced Study of Sustainability) (2018) Case study database

van Oudenhoven FJW, Mijatović D, Eyzaguirre PB (2010) Bridging managed and natural landscapes: the role of traditional (agri)culture in maintaining the diversity and resilience of social-ecological systems. In: Bélair C, Ichikawa K, Wong BYL, Mulongoy KJ (Eds) Sustainable use of biological diversity in socio-ecological production landscapes: background to the 'Satoyama Initiative for the benefit of biodiversity and human well-being.' Secretariat of the Convention on Biological Diversity. Montreal. Technical Series no. 52, pp 8–21

Wendling LA, Aapo H, zu Castell-Rüdenhausen M, Hukkalainen M, Airaksinen M (2018) Benchmarking nature-based solution and smart city assessment schemes against the sustainable development goal indicator framework. Front Environ Sci 6:69

Yiping L (2011) Productive bamboo landscapes of Western Zhejiang. Case studies, International Partnership for the Satoyama Initiative. https://satoyama-initiative.org/casestudies.

Yu H, Verburg PH, Liu L, Eitelberg DA (2016) Spatial analysis of cultural heritage landscapes in Rural China: land use change and its risks for conservation. Environ Manag 57(6):1304–1318

Yun-Ju F, Bo-Wen S, Chuan-Kai H, Wen-Tsui L (2015). Facilitating biological and freshwater resource conservation by agricultural activities at Gongliao-Hoho-Terraced-Paddy-Fields, Taiwan. Case studies, International Partnership for the Satoyama Initiative. https://satoyama-initiative.org/casestudies

Chapter 8
Mapping the Current Understanding of Biodiversity Science–Policy Interfaces

Ikuko Matsumoto, Yasuo Takahashi, André Mader, Brian Johnson, Federico Lopez-Casero, Masayuki Kawai, Kazuo Matsushita, and Sana Okayasu

Abstract This chapter contributes to improve an understanding of the effectiveness of different biodiversity science–policy interfaces (SPIs), which play a vital role in navigating policies and actions with sound evidence base. The single comprehensive study that was found to exist, assessed SPIs in terms of their 'features'—goals, structure, process, outputs and outcomes. We conducted a renewed systematic review of 96 SPI studies in terms of these features, but separating outcomes, as a proxy for effectiveness, from other features. Outcomes were considered in terms of their perceived credibility, relevance and legitimacy. SPI studies were found to focus mostly on global scale SPIs, followed by national and regional scale SPIs and few at subnational or local scale. The global emphasis is largely explained by the numerous studies that focused on the Intergovernmental Platform on Biodiversity and Ecosystem Services (IPBES). Regionally, the vast majority of studies were European, with a severe shortage of studies, and possibly SPIs themselves, in especially the developing world. Communication at the science–policy interface was found to occur mostly between academia and governments, who were also found to initiate most communication. Certain themes emerged across the different features of effective SPIs, including capacity building, trust building, adaptability and continuity. For inclusive, meaningful and continuous participation in biodiversity SPIs, continuous, scientifically sound and adaptable processes are required. Effective, interdisciplinary SPIs and timely and relevant inputs for policymakers are required

I. Matsumoto · Y. Takahashi (✉) · A. Mader · B. Johnson · F. Lopez-Casero · M. Kawai · S. Okayasu
Institute for Global Environmental Strategies (IGES), Hayama, Kanagawa, Japan
e-mail: yasuo.takahashi@iges.or.jp

K. Matsushita
Institute for Global Environmental Strategies (IGES), Hayama, Kanagawa, Japan

Graduate School of Global Environmental Studies, Kyoto University, Kyoto, Japan

© The Author(s) 2020
O. Saito et al. (eds.), *Managing Socio-ecological Production Landscapes and Seascapes for Sustainable Communities in Asia*, Science for Sustainable Societies, https://doi.org/10.1007/978-981-15-1133-2_8

to ensure more dynamic, iterative and collaborative interactions between policy-makers and other actors.

Keywords Natural capital · Ecosystem services · SPI · Science–policy dialogue · Transdisciplinary · Environmental policy · Biodiversity policy · Stakeholder participation · CRELE · IPBES · Knowledge holder · Policy impact · Trust building

8.1 Introduction

A strong interface between science and policy is essential for the effective conservation and management of biodiversity. Science–policy interfaces (SPIs) can generally be defined as social processes encompassing the relations between scientists and actors in the policy process (van den Hove 2007) and can take the form of organizations, initiatives, or projects operating at the boundary between science and policy. SPIs aim to enrich decision-making, improve understanding of problems, and eventually produce well-informed policy and/or behavioural changes as outcomes (Sarkki et al. 2015). The perceived importance of SPIs for biodiversity-related decision-making has been demonstrated by the formulation of the Intergovernmental Science-Policy Platform on Biodiversity and Ecosystem Services (IPBES). The purpose of IPBES is to establish a continuous dialogue between decision-makers, scientists and a wide range of knowledge holders for a more robust SPI on biodiversity and ecosystem services, based in large part on a series of comprehensive assessments on pressing conservation issues (Larigauderie and Mooney 2010a).

There is general consensus on the need for good biodiversity science to inform policy decisions (Millennium Ecosystem Assessment [MA] 2005; Sutherland et al. 2004), and a number of approaches to synthesize scientific knowledge have been established (Pullin and Stewart 2006; Pullin et al. 2009; Sutherland et al. 2014). A number of institutions and processes aim to bring this knowledge to policy processes but, in practice, often fail to produce meaningful policy outcomes (Koetz et al. 2008; Can et al. 2009). Furthermore, they often fail to include the full range of existing knowledge and knowledge holders. Consequently, networking and communication components among different stakeholders are not adequately reflected in many existing SPIs related to biodiversity and other scientific fields (Nesshöver et al. 2016).

A number of studies have demonstrated that effective SPIs do not consist of simple knowledge transfer. The linear model of academics providing scientific advice to governments for policymaking has been rejected from both the perspectives of science studies and policy analysis (van Eeten 1999). Instead, reciprocal (rather than unidirectional) relationships are preferable (Weingart 1999). Scientific knowledge is commonly viewed as information that is useful for problem-solving, but this is only one of a series of different possible uses of science (Roqueplo 1995). Science is a source of legitimacy in the policy process, not only for developing new policies, but also for delaying or avoiding action and for justifying unpopular decisions (Boehmer-Christiansen 1995). In many cases, scientific knowledge is unused or under-used in the policy process (Hisschemöller et al. 2001). Even if particular scientific evidence is used for policymaking, it may remain unclear why it was used while other knowledge is ignored. Scientific rationalization has become an important factor in policymaking, but the decision to connect a policy decision to scientific evidence (and the way in which this is done) depends on political, not academic, factors. Organizing successful SPIs requires some understanding of how the policy process works and how scientific expertise is typically treated in the policy process (Engels 2005).

SPIs have been studied at various geographical scales. Borie and Hulme (2015) looked at the global level with the debate among IPBES experts terminology to include in the IPBES conceptual framework. The key solution was the presence of mediating experts, who finally facilitated the inclusion of both competing terms. At the regional level, Santos and Pierce (2015) reviewed the early implementation of the EU Marine Strategy Framework Directive, focusing on its cetacean biodiversity component. They identified the potential solutions including securing funding for monitoring, reconciling conservation objectives with the needs of other marine/maritime sectors, and clarifying governance structure. At the national level, López-Rodríguez et al. (2015) examined the establishment of an SPI between scientists and policymakers to understand the major environmental problems and priorities in southeastern Spanish drylands. Possible solutions identified for facilitating/operationalizing SPIs included matching different professional groups with concrete problems in their own work fields, using graphical tools to facilitate mutual understanding, clarifying the roles involved in the problem-solving, and promoting a culture of shared responsibility for implementing collaborative actions to solve environmental problem(s). At the subnational level, Chaves et al. (2015) relayed some lessons from a new environmental restoration policy in São Paulo State, Brazil. The study noted that the main solution for effective restoration policymaking is to gain cooperation among scientists, policymakers, and experienced practitioners in identifying appropriate and user-friendly ecological indicators and associated protocols for monitoring and evaluation. These studies suggest a need to share clear visions of SPIs (Santos and Pierce 2015); resource allocation and good governance for SPIs (Santos and Pierce 2015); engagement of different stakeholders and clarification of each of the roles (López-Rodríguez et al. 2015); and collaboration, trust building, capacity building, and conflict management among different stakeholders (Borie and Hulme 2015), in order to improve biodiversity SPIs.

This chapter presents a systematic review of literature on existing SPIs, identifying challenges and possible solutions to effective SPI implementation. This was done in the context of key SPI features—goals, structure, process and outputs, and their policy outcomes (Young et al. 2013a). These SPI features are borrowed from the SPIRAL project (Science-Policy Interfaces for Biodiversity: Research, Action and Learning), an interdisciplinary research project that studied biodiversity SPIs in an attempt to improve the conservation and sustainable use of biodiversity funded under the EU 7th Framework Programme. SPI **goals** are central to understanding how and why an SPI operates, why people participate, and play a strong role in setting the foundations of credibility, relevance, and legitimacy. SPI **structure** describes how SPIs are set up and the constraints within which the processes are defined. SPI **processes** define the way in which the key functions are actually carried out. SPI **outputs** can be characterized by a set of features describing how they are prepared and presented. SPI **outcomes** are the learning, behavioural, and policy changes that SPIs foster. Table 8.1 provides more a further breakdown of these categories. SPIRAL evaluated SPIs based on their perceived credibility, relevance, and legitimacy (CRELE) (Young et al. 2013a; Nesshöver et al. 2016). In this context, **credibility** is defined as 'the perceived quality, validity and scientific adequacy of people, processes and knowledge exchange at the interface'; **relevance** is 'the perception of the usefulness of the knowledge brokered in the SPI, how closely it related to the needs of policy and society, and how responsive the SPI processes are to these changing needs'; and **legitimacy** is 'the perceived fairness and balance of the SPI process' (Young et al. 2013a).

8.2   Methodology

We searched the peer-reviewed literature on biodiversity-relevant SPIs using a search string of '(science-policy OR policy-science) AND biodiversity' on Scopus (https://www.scopus.com/home.uri), one of the largest databases of peer-reviewed literature. The resulting papers' titles, keywords, and abstracts were screened to identify those on *creating or analysing* SPIs. From these, we extracted information on the relevant SPI study including its location, its spatial scale (subnational, national, regional, or global), the associated key challenges and possible solutions identified in the SPI process, and its outcomes.

We assessed each article's analysis of the relevant SPI according to the SPI features identified by the SPIRAL framework on goals, structure, processes, outputs, and outcomes, as listed in Table 8.1. Our assumption is that the outcomes of an SPI are affected by the other four SPI features, but also by external factors (e.g., political climate and pure chance). For this reason, SPI outcomes may not directly reflect the SPI design/operation choices in the same way as other features (Young et al. 2013b). Most studies evaluated the other four features in terms of their perceived credibility, relevance, and legitimacy (CRELE) as a proxy for evaluating the SPI outcomes.

The literature review under this study focused therefore on identifying the major challenges faced by each of the SPI features (goal, structure, process, and output), as well as the possible solutions to these challenges for each of the features. We also assessed how goal, structure, process, and output contribute to better SPI outcomes, through a systematic review of the studies that analysed their causal link. This was

Table 8.1   Key features of SPIs (adapted from Young et al. 2013a, b)

Feature	Sub-feature	Characteristics
Goals	Vision	Clarity, scope, and transparency of the vision and objective of SPI
	Drivers	Demand-pull from policy, mandates, supply-driven promotion of research, emerging issues
Structure	Independence	Freedom from external control, neutrality or bias in position, range of membership
	Participation	Range of relevant expertise and interests included, competence of participants, openness to new participants
	Resources	Financial resources, human resources (e.g., leadership, champions, ambassadors, translators), networks, time
Processes	Horizon scanning	Procedures to anticipate science, technology, policy, and societal developments
	Continuity	Continuity of SPI work on the same issues; continuity of personnel; iterative processes
	Conflict management	Strategies such as third party facilitation, allowing sufficient time for compromise
	Trust building	Possibilities to participate in discussion, clear procedures, opportunities for informal discussions, transparency about processes and products
	Capacity building	Helping policymakers to understand science and scientists to understand policymakers, building capacities for further SPI work
	Adaptability	Responsiveness to changing contexts, flexibility to change
Outputs	Relevant outputs	Timely in respect to policy needs, accessible, comprehensive, efficient dissemination
	Quality assessment	Processes to ensure quality, comprehensiveness, transparency, robustness, and management of uncertainty
	Translation	Efforts to convey messages across different domains and individuals, and making the message relevant for various audiences
Outcomes	Social learning	SPI participants, audiences, wider public learn and change their thinking about biodiversity
	Behavioural impact	SPI participants, audiences, wider public change behaviour as a result of learning
	Policy impact	SPI information, learning, and associated changes in policymaker behaviour lead to changes in policy
	Biodiversity impact	The above changes lead to changes in drivers and pressures threatening biodiversity, societal responses, and the state of biodiversity

considered an indication of their effectiveness. SPI outcomes were captured by their reported impacts, broadly categorized as social learning impacts, policy impacts, behavioural changes, and biodiversity impacts. The subcategories of SPI features, as listed in Table 8.1, were used as units to analyse impacts.

8.3   Results and Discussion

We identified 178 peer-reviewed articles, published from 1990 to April 6, 2017, with titles, keywords, or abstracts containing our search terms. Ninety-six of these were found to be directly relevant to our review of biodiversity SPIs, which discussed about science policy interface on biodiversity in the articles. As illustrated in Fig. 8.1, the number of these studies has been increasing overall since 2008. This may be due to discussion towards the establishment of IPBES in 2012 and an increasing scholarly interest in SPIs in general for improving environmental policy making.

The subsequent section are based on a systematic literature review of existing studies on biodiversity SPIs, analysing (1) the geographic scales/locations of the SPIs that have been studied, (2) the types of SPI features that have been studied, and (3) the challenges, solutions, and outcomes identified in relation to each SPI feature.

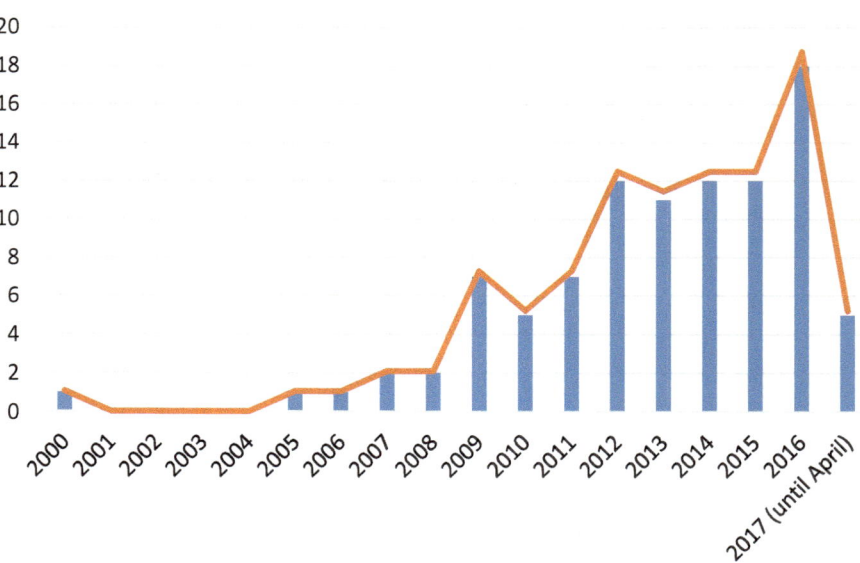

Fig. 8.1 Number of SPI study in each year (2000–2017)

8.3.1 Distribution of SPI Studies

The largest number (36 out of 96) of articles studied global-level SPIs, mostly related to IPBES, while subnational/local and cross-scale SPIs received the least research attention (Fig. 8.2). Rather than reflecting the existence of only a limited number of subnational/local SPIs, this indicates a shortage of studies focusing on the numerous SPIs related to local biodiversity conservation plans and policies. In terms of where the regional-, national-, and subnational/local level SPIs were studied (37 in total), the majority focused on SPIs in Europe (22) and North America (5), while comparatively few studies focused on Asia (3), Oceania (2), Latin America (2), and Africa (1) (Fig. 8.3). Forty-four out of the 96 SPIs were facilitated by government or by government and academia (10), while only 13 were facilitated by academia alone (Fig. 8.4). Most papers (64) involved SPIs with two-way communication between scientists and policymakers, typically with multiple rounds of presentation and feedback. The second most common means of communication was a linear style of one-way communication from scientists to policymakers or vice-versa (12) (Fig. 8.5). Some SPIs used both collaborative and linear means of communication.

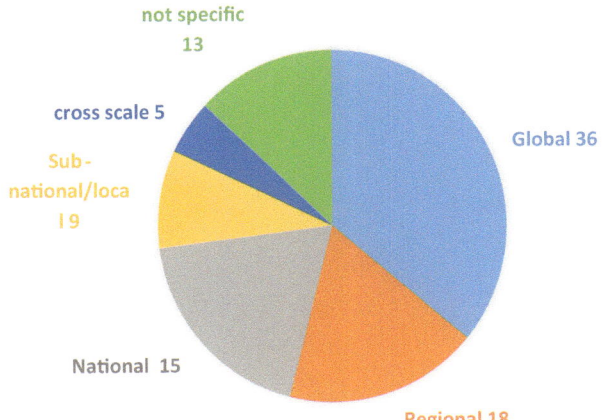

Fig. 8.2 Geographical scale of SPI studies

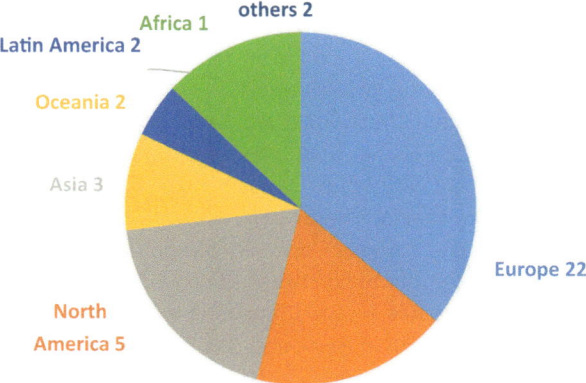

Fig. 8.3 Regional balance of SPI studies

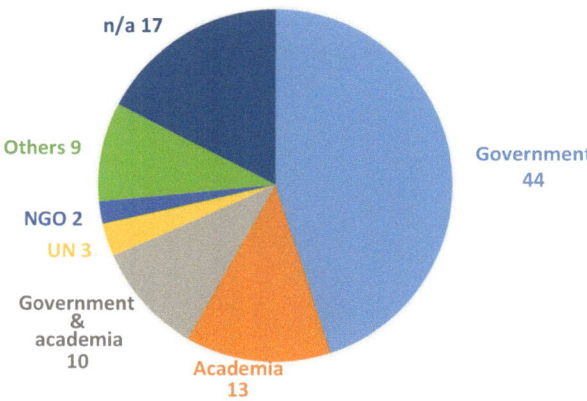

Fig. 8.4 Facilitators of SPI

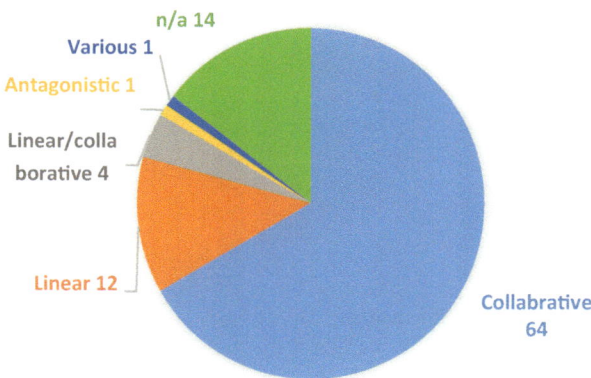

Fig. 8.5 Way of communication in SPI

8.3.2   SPI Features

8.3.2.1   Overview

Challenges and Possible Solutions

Of the 96 articles relevant to our review, 77 discussed the challenges faced, and the possible solutions provided, by SPIs. Some of these identified more than one key feature of the SPI studied; therefore, the total number of features mentioned does not match the total number of articles reviewed. Other 19 articles did not discuss any particular features, challenges, and possible solutions of SPIs. Of the 77 articles analysed, 45 articles discussed challenges and possible solutions in the SPI *process*, specifically in terms of capacity building (18 studies), trust building (16), adaptability (12), continuity (10), horizon scanning (9), and conflict management (5) (Fig. 8.6). Challenges and possible solutions regarding SPI *structure* were discussed in 34 articles, mostly in terms of participation (29). Nineteen articles discussed SPI *output* challenges and possible solutions, with relevance of outputs (12) being the most frequently assessed output component. Challenges and possible solutions related to SPI *goals* were least frequently covered in the literature (only 11 studies), especially in terms of the drivers of the SPI development (e.g., whether it was set up due to policy demand, research interest, or new emerging issues) being evaluated most often (7 studies).

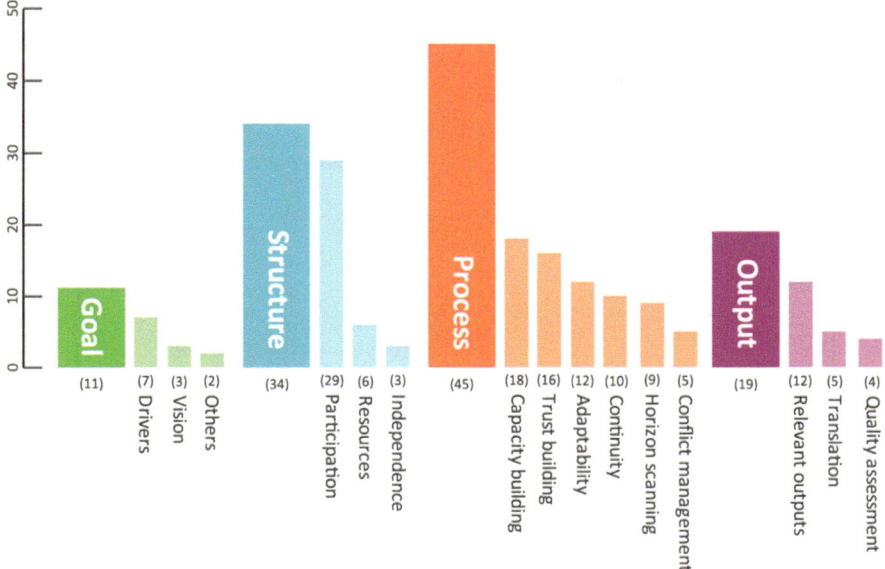

Fig. 8.6   Key SPI features recognized in the articles

Table 8.2 Key challenges and possible solutions identified in the reviewed articles

	Challenges	Possible solutions
Goal	• Identification of key research topic	• Joint formulation of research and policy between researchers and policymakers
	• Goals and objectives of SPI is not clear	• Developing and adjusting clear goal and priority of SPI for participants
Structural	• Assembling a range of knowledge holders and experts relevant to topics	• Formation of SPIs with transparent and open structures
	• High level of complexity of decision-making	• More engagement with social sciences
	• Need to ensure a sound scientific basis of SPI	• Collaborative interdisciplinary teams and involve scientists, policymakers, legal experts, and practitioners from various fields/sectors on board
	• Fragmentation of group of interests of the members involved in SPI	• Establishment of a discussion platform among different stakeholders
		• Putting in place structures and incentive schemes that support long-term interactive dialogue
Process	• Overcoming silos between decision-makers and scientists	• Adequate capacity building for both scientists and policymakers to understand the different processes in which each of them work
	• Appropriate handling of socio-ecological complexity and political dimensions	• More engagement with social sciences
	• Timely provision of consolidated view for decision-making	• Enhancing national level of capacity including data collection and technical skill
	• Better communication between policymakers and scientists and addressing or communicating the uncertainty of science	• Engagement of policymakers in research projects
	• Striking an appropriate balance between scientific complexity and over-simplification	
	Improvement of data collection and use	
	• Lack of common language or philosophies between scientists and policymakers	
Outputs	• Making scientific output policy relevant	• Integrating knowledge more with social science including socioeconomic impacts
	• Transforming knowledge between different communities	• Production of highly relevant outputs of SPIs
	• Need to strengthen scientific basis	

Many SPI studies (29) mentioned participation as a key challenge of SPI structure. Capacity building (18 studies) and trust building (16 studies) were also described as key challenges. Table 8.2 summarizes the common challenges and possible solutions identified in the reviewed papers.

Outcomes

Of the 96 relevant articles, we identified 42 that examined how the goals, structure, process, and output of existing SPIs affected the wider outcomes of the SPI process. In the 42 articles, we identified 92 cases in which outcomes were reported. Among the four SPI features, the SPI process was by far the most studied (52 cases), followed by structure and output. These results were quite similar to those related to SPI challenges and solutions, with process and structure being among the most discussed in relation to outcomes. The relationship between SPI goals and outcomes was the least studied (only 6 cases). As for the types of outcomes investigated, social learning and policy impacts were most studied (Table 8.3). Table 8.4 summarizes different efforts and tools falling under the four SPI feature categories and their outcomes, drawn from our systematic review. These are described more in detail in the following four sections.

8.3.2.2 SPI Goal

Some of the common challenges to achieving the goals of SPIs included the identification of relevant research topics (Pullin et al. 2009; Vohland et al. 2011; Sarkki et al. 2013) as well as a lack of clarity about what these goals and objectives should be (Chapple et al. 2011; Kim et al. 2016). The most common possible solutions identified for overcoming these common challenges included joint formulation of research that would produce science to inform policy, by scientists and policymakers (Noss et al. 2009; Pullin et al. 2009; Ferreira et al. 2012; Sarkki et al. 2013; Young et al. 2014; Chaves et al. 2015; Nesshöver et al. 2016), and developing and adjusting clear goals and priorities of SPIs among different stakeholders at the initial stage of the SPI formulation (Kim et al. 2016). In a survey of the scientific community on the need and possible options for a science–policy platform, many

Table 8.3 Number of the cases of causal link between the four SPI features and outcomes

SPI feature category	SPI outcome subcategory				
	Social learning	Policy impact	Biodiversity impact	Others	Subtotal
Goal	1	4		1	6
Structure	8	5		4	17
Process	23	22	1	6	52
Outputs	3	7	1	6	17
Total	35	38	2	17	92

Table 8.4 Summary of causal link between the four SPI features and their outcomes

SPI features	SPI approaches and tools	Outcomes
Goal	• SPI with clear political mandate	• Higher likelihood of policy uptake
	• Pluralistic and relatively open political structure	• Enable science to frame problems for policies
	• Balance between policy-led and science-led approaches	• Science helps form new priorities, and science responds to imminent policy needs
	• Scientists' and policymakers' joint effort	• Identify priority issues, deliver consolidated knowledge to support policies, and identify research gaps to address emerging issues
Structure	• Balanced participation across space and disciplines	• Sound spatial and disciplinary representation in SPI deliverables
	• Boundary object/participatory assessment	• Trigger diverse stakeholders to work collectively and to share understanding
	• Transdisciplinary institution	• Use of credible scientific results in policies
Process	• Clear protocols and higher transparency	• Create long-lasting mutual trust and learning environment
	• Regular face-to face interactions between science and policy; inclusion of policymakers in research projects	• Enhance mutual understanding between policymakers and researchers
	• Acknowledge and spur the enthusiasm of diverse participants	• Integrate different forms of knowledge and use them in decision-making
	• Adaptive learning by doing framework	• Use of research results in management
Output	• Strengthen social science engagement	• Respond to policymakers' need for identifying effective policies
	• Knowledge synthesis, e.g., policy briefs, white paper, database, red-listing	• Policy changes

respondents considered decision-making (i.e., policymaking) to be complex, iterative, and often selective in the information used. The authors concluded that joint formulation of policy would be preferable (Young et al. 2014).

In terms of the contributions of SPI goals to outcomes, most studies (four of five) investigated how the goal features were related to policy impacts. Sarkki et al. (2013) found a higher likelihood of policy uptake of the findings and recommendations from SPIs with a predetermined political mandate, referring to experience from the Intergovernmental Panel on Climate Change (IPCC). Conversely, pluralistic and relatively open political structures and processes were found to enable scientists to better identify and prioritize problems for policies (Tzankova 2017). Thus, both policy-led and science-led approaches have merits, where science can help identify new priorities while responding to existing policy needs (Sarkki et al.

2013). This is well supported by López-Rodríguez et al. (2015) and Nesshöver et al. (2016), who noted the contribution of scientists' and policymakers' joint efforts to identify priority environmental issues, delivering a consolidated body of scientific knowledge to support relevant policies, as well as to identify research gaps to address emerging issues. This also applies to ecosystem management. Drawing on their experience with scientists' engagement in the management of Greater Blue Mountains World Heritage Area in Australia, Chapple et al. (2011) emphasized the importance of the collaboration and information flow between scientists and managers to collectively define problems and management objectives that guide research directions and uptake.

8.3.2.3 SPI Structure

The most common challenges to structuring SPIs included assembling a range of knowledge holders and experts relevant to topics (Ferreira et al. 2012; Plant and Ryan 2013; Spranger et al. 2014; Schewenius et al. 2014; Hauck et al. 2014; Walther et al. 2016); the high level of complexity of decision-making processes (Young et al. 2014; Tzankova 2017); and the need to ensure a sound scientific basis of the SPIs. A lack of incentives for scientists and policymakers to participate in SPIs (Granjou and Mauz 2012; Sarkki et al. 2013) and fragmentation of interests of the members involved in the SPIs (Gustafsson and Lidskog 2013; Hauck et al. 2014; Arpin et al. 2016) constitute further challenges.

In terms of solutions to these problems, the formulation of SPIs with transparent and open structures was frequently identified as a solution. For example, Arpin et al. (2016) found that the major challenges in the process of establishing IPBES were handling the fragmentation and plasticity of the group of interest involved in the institutionalization process, and the 'exercise of an art of having everybody on board through techniques of inclusiveness' was a key to success. Many studies observed that, in order to tackle complex and multidimensional issues of biodiversity, it is vital to have collaborative interdisciplinary teams and to involve scientists, policymakers, legal experts, and practitioners from various fields/sectors (Srebotnjak 2007; Koetz et al. 2008; Arts and Buizer 2009; Mishra et al. 2009; Blythe and Dadi 2012; Ferreira et al. 2012; Kueffer et al. 2012; Paloniemi et al. 2012; Giakoumi et al. 2012; Ardoin and Heimlich 2013; Gustafsson and Lidskog 2013; Keune et al. 2013; Young et al. 2014; Hauck et al. 2014; Chaves et al. 2015; Sarkki et al. 2015; Andaloro et al. 2016; Arpin et al. 2016; Kovács and Pataki 2016; Walther et al. 2016). Kueffer et al. (2012), noting the complexity of problems, and impartiality of expertise and salience of knowledge which impede effective research for sustainable development, found that one solution is to conduct research in interdisciplinary teams, forming research partnerships with actors and experts from outside academia, and framing research questions with the aim of solving specific problems. In order to do so, Seddon et al. (2016) suggested that ecologists and conservation biologists need to engage much more strongly with, and draw on, the social sciences as well as the humanities. It was also considered critical to establish a discussion platform among different stakeholders (Sommerwerk et al. 2010b; Cil and Jones-Walters 2011; Thomas et al. 2012; Mace et al. 2013; Spranger et al. 2014; Schewenius et al. 2014; Garibaldi et al. 2017). Putting in place structures and incentive schemes that support long-term interactive dialogue, such as new network opportunities, recognition in an academic society, access to funding and others (Granjou and Mauz 2012; Young et al. 2014; Hauck et al. 2014; Carmen et al. 2015; Santos and Pierce 2015; Sarkki et al. 2015; Nesshöver et al. 2016) was another possible solution to address these challenges.

In order to address these challenges and secure sound participation among different stakeholders in long term, trust building in the SPI process is important to facilitate engagement with social scientists, multiple sectors of governments, practitioners, private sectors, and others. To ensure participation from local and indigenous communities, capacity building and different communicative forms are vital at the same time. Kim et al. (2016) stated that increased participation, per se, does not guarantee the achievement of ethical-moral imperatives (people should have a say in decisions affecting them) or instrumental outcomes such as improving people's ownership and acceptance. To address structural challenges of SPIs, they also pointed to the question of *how* the process was conducted as also being important. And it is affected by institutional culture, transparency, flexibility, and capability for implementation. Mielke et al. (2017) evaluated stakeholder involvement practices in science and concluded that 'more conceptual exchange between practitioners, as well as more qualitative research on the concepts behind practices, is needed to better understand the stakeholder–scientist nexus'. Active engagement of stakeholders with a range of relevant expertise and interest will help an SPI to better handle the socio-ecological complexity and political dimensions of biodiversity-related policy-making. Further, improvement of SPI processes including trust building, continuity, capacity building, and adaptability will also lead to more robust SPI structure (e.g., resulting in more active participation within the SPI). This demonstrates the dynamic relationship between 'structure' and 'process' of SPIs. So, to promote more meaningful and continuous participation in biodiversity SPIs and better SPI structure, it is not enough to invite experts and stakeholders from different sectors to participate in SPIs, but also to secure continuous, trusted, and adaptable SPI processes.

In terms of how the structure of SPIs can contribute to specific outcomes, most studies focused on their social learning impacts (8 out of 17) and policy impacts (5 out of 17). As for social learning, participatory assessment, e.g., biodiversity assessment that involves various stakeholders including scientists and policymakers, can be used to generate comprehensive evidence and underpin shared understanding among stakeholders (Garibaldi et al. 2017). Sarkki et al. (2013) reported that the participation of governments in the IPCC decision-making process increased their likelihood of referencing the IPCC assessments in their policies. Regarding policy impacts, Kovács and Pataki (2016), drawing on their observation of the early-stage development of IPBES, highlighted the need for diverse and balanced participation of experts across regions and countries to ensure the representation of place-specific knowledge in global- and regional-level assessments. Balanced participation was also found to enhance legitimacy in priority setting (Kim et al. 2016). Diverse participation allows for bridging of knowledge and skills between experts and public beyond traditional boundaries (Carmen et al. 2015; Andaloro et al. 2016). Transdisciplinary SPIs at regional, national, and local levels saw several cases of success in policy uptake. These included the use of scientific results to define the limits of emission values, best available techniques, and economic instruments under the Convention on Long-Range Transboundary Air Pollution (CLRTAP) (Spranger et al. 2014); the development of England's national biodiversity strategy building on the national ecosystem assessment report delivered by a team of multi-disciplinary experts and policymakers (Watson 2012); and the integration of science–policy activities under the International Commission for the Protection of the Danube River (ICPDR) (Sommerwerk et al. 2010a). Problem-oriented and interdisciplinary research and partnership were found to drive transitional change of academic culture (Kueffer et al. 2012).

8.3.2.4 SPI Process

Overcoming silos between decision-makers and scientists (Tinch et al. 2016; Carmen et al. 2015; Lidskog 2014; Sanguinetti et al. 2014; Sarkki et al. 2013; Ruckelshaus et al. 2015; Aslaksen et al. 2012; Koetz et al. 2012; Naylor et al. 2012; Noss et al. 2009; Srebotnjak 2007) and timely provisioning of consolidated views for decision-making (Larigauderie and Mooney 2010b; Thomas et al. 2012; Carmen et al. 2015; Nesshöver et al. 2016) were identified as key challenges to the process of developing and maintaining SPIs. Many articles also emphasized the need for interdisciplinary SPIs to develop policies that can take into account the complexity and interconnectedness of social and ecological systems (Arts and Buizer 2009; Mishra et al. 2009; Pullin et al. 2009; Van Haastrecht and Toonen 2011; Blythe and Dadi 2012; Kueffer et al. 2012; Paloniemi et al. 2012; Keune et al. 2013; Young et al. 2014; Hauck et al. 2014; Sarkki et al. 2015; Raina and Dey 2015; Seddon et al. 2016; Chazdon et al. 2017).

One potential solution to these challenges, which was identified in several past studies on individual SPIs, could be to put in place incentives for scientists and policymakers to support their long-term, interactive dialogue as well as the collaboration of diverse stakeholders and knowledge holders. Some authors noted that contribution to better decision-making required better communication between policymakers and scientists and addressing or communicating the uncertainty of science (Opdam et al. 2009; Rodela et al. 2015; Balian et al. 2016). At the same time, the need was recognized to strike an appropriate balance between scientific complexity on the one hand and over-simplification on the other (Sarkki et al. 2013; Balian et al. 2016). Improvement of data collection and use (Ruckelshaus et al. 2015; Stephenson et al. 2015) and lack of common language or philosophies between scientists and policymakers (Borie and Hulme 2015; Rodriguez et al. 2015; Sarkki et al. 2015; Gigante et al. 2016; Tremblay et al. 2016) were also singled out as means for a better decision-making process between these two groups.

Adequate capacity building for both scientists and policymakers to understand the respective processes in which they work was stated as a key SPI process in 18 reviewed articles. For instance, discussing biodiversity data for decision-making in Africa, Stephenson et al. (2015) stressed the importance of building capacity for data collection, using tools, guidelines, and communities on biodiversity planning and monitoring. In order to promote interaction between scientists and decision-makers to improve mutual understanding in Africa, they also mentioned the need for the improvement of national, international, and cross-sectoral collaboration for biodiversity data management, and the production and use of more data-derived products that encourage data use. Ruckelshaus et al. (2015) pointed out the importance of training local experts in the use of different approaches and tools for building local capacity, ownership, trust, and long-term success. Neßhöver et al. (2013) found that, if policy requires a broad foundation and exhaustive interdisciplinary synthesis, broad assessments such as Millennium Ecosystem Assessment (MA) or The Economics of Ecosystem *and* Biodiversity (*TEEB*) would be more effective in the engagement of policymakers.

Trust building was also frequently identified (in 16 articles) as being a relevant solution to address the challenges in SPI processes, and it is closely related to capacity building. For example, to identify and overcome the numerous social, cultural, and political obstacles to effective transition of policy into action and financial resources that benefit biodiversity, Seddon et al. (2016) stated that ecologists and conservation biologists need to engage much more strongly with, and draw on, the social sciences and the humanities.

In terms of the contributions of SPI processes to outcomes, most studies described the social learning (23 of 52) and policy impacts (22 of 52) in an inseparable continuum. Tinch et al. (2016) found that long-lasting mutual trust and a learning environment were vital to generate positive SPI outcomes including social learning and policy impacts, drawing from a review of ten SPIs at national, regional, and global levels. Clear procedural protocols and higher transparency in SPI process were found to also enhance mutual trust (Kim et al. 2016). Regular face-to-face interactions between scientists and policymakers (Balian et al. 2016), as well as their exchange in the upstream of the research project design process (Neßhöver et al. 2013), can enhance mutual understanding between policymakers and researchers and accelerate the flow of scientific knowledge into policies and practices, and the inclusion of policy perspectives into research projects. Such a reciprocal and iterative process helps policymakers understand and deal with uncertainties, and strengthen learning in and policy relevance of SPI (Sarkki et al. 2013; Balian et al. 2016). In doing so, it was recommended to acknowledge and spur the enthusiasm of various participants to bring different forms of knowledge together and to integrate knowledge in decision-making (Carmen et al. 2015). Ruckelshaus et al. (2015) suggested the need for focused capacity building for local experts on the approaches and tools to enhance local capacity, ownership, and trust, which helps integrate local values in biodiversity planning. Sarkki et al. (2013), on the other hand, were of the opinion that scientists need to be better aware of the cycle of the policy process that they intend to influence. All in all, continuous interaction between scientists and policymakers from an earlier stage supports more targeted and timely inputs of quality knowledge from scientists in policy cycle (Sarkki et al. 2013; Balian et al. 2016), and an adaptive process would enable appropriate response to changing policy needs and to help shape next generation of policy questions (Sarkki et al. 2013). In policy implementation, an adaptive 'learning by doing' framework was considered to enhance the use of research results (Chapple et al. 2011).

8.3.2.5   SPI Output

Common challenges related to the outputs of SPIs included making scientific outputs policy relevant (Mishra et al. 2009; Vohland et al. 2011; Balian et al. 2016; Donohue et al. 2016; Nesshöver et al. 2016) and an inadequate scientific basis of outputs for policymaking (Koetz et al. 2008; Donohue et al. 2016). The production of highly relevant outputs of SPIs was most frequently cited as a solution, with the relevance of the output being enhanced typically through several rounds of communication between scientists and policymakers. For example, given the impact of conventional intensification of agriculture on biodiversity loss and greenhouse gas emission, Garibaldi et al. (2017) stated an urgent need to provide quantitative evidence of simultaneous ecological and socioeconomic impacts across the globe by alternative agriculture approaches to direct science–policy initiatives, such as SDGs and IPBES. They also proposed a participatory assessment framework as one of the possible solutions to close this knowledge gap. In Brazil, facing the knowledge gaps regarding the ecological impacts of agricultural expansion and the general disconnection between ecological science and environmental policy development processes, Joly et al. (2010) stated that the efforts to synthesize data for policymaking and state-level demand were important for the success of biodiversity conservation.

In terms of how outputs contribute to positive outcomes, most studies focused on policy impacts (7 of 17 studies) and social learning (3 of 17 studies), where social learning was described as a process leading to policy impacts. Extended peer-reviews and well-defined quality assessment process were found to enhance the learning of participants and enhance the quality of outputs (Sarkki et al. 2013; Beck 2014). Diverse ways of presenting synthesized knowledge, including policy briefs, are used as a reliable and handy evidence base for policymaking. For a decision on marine management rules, policy briefs, pictures, maps, and figures were found to be efficient translation tools for simplifying message for policymakers (Sarkki et al. 2013). The BIOTA-FAPESP programme on biodiversity conservation research in the state of São Paulo has provided research underpinning of 4 governmental decrees and 11 resolutions through its efforts to synthesize data in response to the public and state's demand (Ferreira et al. 2012). In the United Kingdom, the National Ecosystem Assessment report was referred to in the National Environment White Paper, which was used to develop a national biodiversity strategy (Watson 2012). IUCN's Red List is a good example of a credible quality SPI output which has become frequently referred to in policies as the representation of the state of biodiversity (Gustafsson and Lidskog 2013). Advancing information technologies for knowledge integration, such as database and semantic web technologies, enable ecosystem managers to easily access expert knowledge (Blythe and Dadi 2012).

8.3.2.6 SPI Outcome

We can draw some important findings from the above analysis of the outcomes from each of the four SPI features in terms of how they can contribute to enhancing the credibility, relevance, or legitimacy in SPI. Sarkki et al. (2013) highlighted the potential trade-offs between credibility, relevance, and legitimacy in SPIs. The trade-offs, however, are highly context dependent. Our analysis identified generic approaches and tools to reconcile the trade-offs and enhance synergies between credibility, relevance, and legitimacy in SPI. Under SPI goals, scientists' and policymakers' joint efforts with their appropriate power balance can merit the synergies. As for SPI structure, transboundary institutions that ensure a good representation of policymakers, scientists, and other stakeholders in relevant and diverse sectors and disciplines can contribute to enhancing the synergies. The synergies can also be improved through an SPI process with clear protocols for higher transparency and with a mechanism to enhance the enthusiasm of various participants which will also contribute to building synergies.

8.4   Conclusion

In terms of the geographic scale and locations of the SPIs studied, we found that most were global (mainly IPBES) or regional or national SPIs in Europe or North America. Relatively few studies investigated regional or national SPIs in Asia, Africa, or South America, despite the importance of these regions in terms of biodiversity conservation. Studies focusing on the numerous SPIs related to local biodiversity conservation plans and policies are particularly scarce.

The main challenges and solutions facing SPIs are related to participation, although different terms are used to refer to it in different studies (such as 'joint', 'collaborative', 'participative', and 'involve'). Although participation is classified as a sub-feature of SPI *structure* in Table 8.1, it is a critical component of the other SPI features as well. For example, the joint formulation of research and/or policies was found to be a possible solution to overcome key challenges related to the SPI goals, such as a lack of clarity regarding the goals and objectives or missing identification of relevant research topics. In the context of the SPI structure, participation was found to be a particularly relevant sub-feature. To overcome the existing challenges such as a lack of sound scientific basis, high complexity of decision-making processes, and fragmentation of interests, a key solution proposed in many studies focuses on improving participation by establishing collaborative interdisciplinary and multi-stakeholder structures, such as committees, teams, or partnerships involving scientists, policymakers, legal experts, and practitioners. To be sustainable, however, these participatory structures need to be based on incentive schemes that are able to support the required long-term interactive dialogue to secure continuous, trusted, and adaptable SPI processes. Finally, participatory approaches also constitute possible solutions to challenges faced in the production of SPI *outputs*, by ensuring continuous interaction between scientists, policymakers, and other possible stakeholders to overcome silos and creating participatory assessment frameworks as a possible solution to existing knowledge gaps.

Trust building and capacity building are also important, closely related, possible solutions to existing SPI challenges. Trust building facilitates the engagement of different stakeholders in participatory processes by enhancing the mutual understanding and interaction of scientists and policymakers throughout the stages of setting SPI goals, developing their structures and producing relevant outputs. Flexibility to change and continuity were also identified as relevant sub-features of SPI processes. In this regard, it is vital to ensure more dynamic, iterative, and collaborative interactions between scientists, practitioners, knowledge holders, and policymakers to identify research gaps, consolidate interdisciplinary scientific views, build capacity and long-term trust of organizations, and ultimately develop effective interdisciplinary SPIs that provide timely and relevant outputs to policymakers. Effective instruments for SPIs to deliver credible, relevant, and legitimate outcomes include ensuring a well-defined quality assessment process possibly through extended peer-reviews and the production of a knowledge synthesis that is relevant and handy for knowledge users.

It is important to note that our findings draw on a limited number of studies of a limited number of SPIs. These studies are, furthermore, skewed towards SPIs at global level and/or in Europe and North America. Further studies that empirically assess the features of SPIs and their contributions to outcomes are needed, particularly at underrepresented scales and in underrepresented regions. Further research into how SPI goals and outputs can provide solutions to challenges and lead to positive outcomes is also needed, to develop a more comprehensive choice of approaches that can generate positive outcomes at the science–policy interface.

Acknowledgments This research was supported by the Environment Research and Technology Development Fund (S-15-1(4) Science-Policy interface on Natural Capital and Ecosystem Services in International, Asian and Japanese Contexts; Predicting and Assessing Natural Capital and Ecosystem Services (PANCES)) of the Ministry of the Environment, Japan.

References

Andaloro F, Castriota L, Falautano M et al (2016) Public feedback on early warning initiatives undertaken for hazardous non-indigenous species: the case of Lagocephalus sceleratus from Italian and Maltese waters. Manag Biol Invasions 7(4):313–319. https://doi.org/10.3391/mbi.2016.7.4.01

Ardoin NM, Heimlich JE (2013) Views from the field: conservation educators' and practitioners' perceptions of education as a strategy for achieving conservation outcomes. J Environ Educ 44:97–115. https://doi.org/10.1080/00958964.2012.700963

Arpin I, Barbier M, Ollivier G, Granjou C (2016) Institutional entrepreneurship and techniques of inclusiveness in the creation of the intergovernmental platform on biodiversity and ecosystem services. Ecol Soc 21(4):11. https://doi.org/10.5751/ES-08644-210411

Arts B, Buizer M (2009) Forests, discourses, institutions: a discursive-institutional analysis of global forest governance. For Policy Econ 11(5–6):340–347. https://doi.org/10.1016/j.forpol.2008.10.004

Aslaksen I, Framstad E, Garnåsjordet PA et al (2012) Knowledge gathering and communication on biodiversity: developing the Norwegian Nature Index. Nor Geogr Tidsskr – Nor J Geogr 66:300–308. https://doi.org/10.1080/00291951.2012.744092

Balian EV, Drius L, Eggermont H (2016) Supporting evidence-based policy on biodiversity and ecosystem services: recommendations for effective policy briefs. Evid Policy J Res Debate Pract 12:431–451. https://doi.org/10.1332/174426416X14700777371551

Beck S (2014) Towards a reflexive turn in the governance of global environmental expertise. The cases of the IPCC and the IPBES. GAIA - Ecol Perspect Sci Soc 23(2):80–87(8). https://doi.org/10.14512/gaia.23.2.4

Blythe JN, Dadi U (2012) Knowledge integration as a method to develop capacity for evaluating technical information on biodiversity and ocean currents for integrated coastal management. Environ Sci Pol 19-20:49–58. https://doi.org/10.1016/j.envsci.2012.01.007

Boehmer-Christiansen S (1995) Reflections on scientific advice and EC transboundary pollution policy. Sci Public Policy 22:195–203. https://doi.org/10.1093/spp/22.3.195

Borie M, Hulme M (2015) Framing global biodiversity: IPBES between mother earth and ecosystem services. Environ Sci Pol 54:487–496. https://doi.org/10.1016/j.envsci.2015.05.009

Can OE, Dieterich M, Pullin AS et al (2009) Conservation focus on Europe: major conservation policy issues that need to be informed by conservation science. Conserv Biol 23:818–824. https://doi.org/10.1111/j.1523-1739.2009.01283.x

Carmen E, Nesshöver C, Saarikoski H et al (2015) Creating a biodiversity science community: experiences from a European Network of Knowledge. Environ Sci Pol 54:497–504. https://doi.org/10.1016/j.envsci.2015.03.014

Chapple RS, Ramp D, Bradstock RA et al (2011) Integrating science into management of ecosystems in the greater blue mountains. Environ Manag 48(4):659–674. https://doi.org/10.1007/s00267-011-9721-5

Chaves RB, Durigan G, Brancalion PHS, Aronson J (2015) On the need of legal frameworks for assessing restoration projects success: new perspectives from São Paulo state (Brazil). Restor Ecol 23(6):754–759. https://doi.org/10.1111/rec.12267

Chazdon RL, Brancalion PHS, Lamb D et al (2017) A policy-driven knowledge agenda for global forest and landscape restoration. Conserv Lett 10:125–132. https://doi.org/10.1111/conl.12220

Cil A, Jones-Walters L (2011) Biodiversity action plans as a way towards local sustainable development. Innov Eur J Soc Sci Res 24(4):467–479. https://doi.org/10.1080/13511610.2011.593897

Donohue I, Hillebrand H, Montoya JM et al (2016) Navigating the complexity of ecological stability. Ecol Lett 19:1172–1185. https://doi.org/10.1111/ele.12648

Engels A (2005) The science-policy interface. Integr Assess J Bridg Sci Policy 5:7–26. https://doi.org/10.1016/S1389-9341(03)00041-8

Ferreira J, Pardini R, Metzger JP et al (2012) Towards environmentally sustainable agriculture in Brazil: challenges and opportunities for applied ecological research. J Appl Ecol 49(3):535–541

Garibaldi LA, Gemmill-Herren B, D'Annolfo R et al (2017) Farming approaches for greater biodiversity, livelihoods, and food security. Trends Ecol Evol 32:68–80. https://doi.org/10.1016/j.tree.2016.10.001

Giakoumi S, Mazor T, Fraschetti S et al (2012) Advancing marine conservation planning in the Mediterranean Sea. Rev Fish Biol Fish 22:943–949. https://doi.org/10.1007/s11160-012-9272-8

Gigante D, Attorre F, Venanzoni R et al (2016) A methodological protocol for Annex I habitats monitoring: the contribution of vegetation science. Plant Sociol 53:77–78. https://doi.org/10.7338/pls2016532/06

Granjou C, Mauz I (2012) Expert activities as part of research work: the example of biodiversity studies. Sci Technol Stud 25:5–22

Gustafsson KM, Lidskog R (2013) Boundary work, hybrid practices, and portable representations: an analysis of global and national coproductions of red lists. Nat Cult 8:30–52. https://doi.org/10.3167/nc.2013.080103

Hauck J, Görg C, Werner A et al (2014) Transdisciplinary enrichment of a linear research process: experiences gathered from a research project supporting the European Biodiversity Strategy to 2020. Interdiscip Sci Rev 39:376–391. https://doi.org/10.1179/0308018814Z.00000000098

Hisschemöller M, Dunn WN, Hoppe R, Ravetz J (2001) Knowledge, power and participation in enviornmental policy analysis. Transaction Pulishers, New Brunswick

Joly CA, Rodrigues RR, Metzger JP, Haddad CFB, Verdade LM, Oliveira MC, Bolzani VS (2010) Biodiversity conservation research, training, and policy in São Paulo. Science 328(5984):1358–1359

Keune H, Kretsch C, De Blust G et al (2013) Science–policy challenges for biodiversity, public health and urbanization: examples from Belgium. Environ Res Lett 8:025015. https://doi.org/10.1088/1748-9326/8/2/025015

Kim MK, Evans L, Scherl LM, Marsh H (2016) Applying governance principles to systematic conservation decision-making in Queensland. Environ Policy Gov 26(6):452–467. https://doi.org/10.1002/eet.1731

Koetz T, Bridgewater P, van den Hove S, Siebenhüner B (2008) The role of the subsidiary body on scientific, technical and technological advice to the convention on biological diversity as science–policy interface. Environ Sci Pol 11:505–516. https://doi.org/10.1016/j.envsci.2008.05.001

Koetz T, Farrell KN, Bridgewater P (2012) Building better science-policy interfaces for international environmental governance: assessing potential within the intergovernmental platform for biodiversity and ecosystem services. Int Environ Agreements Polit Law Econ 12(1):1–21. https://doi.org/10.1007/s10784-011-9152-z

Kovács EK, Pataki G (2016) The participation of experts and knowledges in the Intergovernmental Platform on Biodiversity and Ecosystem Services (IPBES). Environ Sci Pol 57:131–139. https://doi.org/10.1016/j.envsci.2015.12.007

Kueffer C, Underwood E, Hadorn GH et al (2012) Enabling effective problem-oriented research for sustainable development. Ecol Soc 17(4):8

Larigauderie A, Mooney HA (2010a) The International Year of Biodiversity: an opportunity to strengthen the science-policy interface for biodiversity and ecosystem services. Curr Opin Environ Sustain 2:1–2. https://doi.org/10.1016/j.cosust.2010.04.001

Larigauderie A, Mooney HA (2010b) The Intergovernmental Science-Policy Platform on Biodiversity and Ecosystem Services: moving a step closer to an IPCC-like mechanism for biodiversity. Curr Opin Environ Sustain 2:9–14. https://doi.org/10.1016/j.cosust.2010.02.006

Lidskog R (2014) Representing and regulating nature: boundary organisations, portable representations, and the science-policy interface. Env Polit 23(4):670–687. https://doi.org/10.1080/09644016.2013.898820

López-Rodríguez MD, Castro AJ, Castro H et al (2015) Science-policy interface for addressing environmental problems in arid Spain. Environ Sci Pol 50:1–14. https://doi.org/10.1016/j.envsci.2015.01.013

Millennium Ecosystem Assessment [MA] (2005) Ecosystems and human well-being: synthesis. Island Press, Washington, DC

Mace GM, Perrings C, Le Prestre P et al (2013) Science to policy linkages for the post-2010 biodiversity targets. In: Biodiversity monitoring and conservation: bridging the gap between global commitment and local action. Wiley-Blackwell, Hoboken, NJ, pp 291–310

Mielke J, Vermaßen H, Ellenbeck S (2017) Ideals, practices, and future prospects of stakeholder involvement in sustainability science. Proc Natl Acad Sci U S A 114:E10648–E10657. https://doi.org/10.1073/pnas.1706085114

Mishra BK, Badola R, Bhardwaj AK (2009) Social issues and concerns in biodiversity conservation: experiences from wildlife protected areas in India. Trop Ecol 50:147–161

Naylor LA, Coombes MA, Venn O et al (2012) Facilitating ecological enhancement of coastal infrastructure: the role of policy, people and planning. Environ Sci Pol 22:36–46. https://doi.org/10.1016/j.envsci.2012.05.002

Neßhöver C, Timaeus J, Wittmer H et al (2013) Improving the science-policy interface of biodiversity research projects. GAIA - Ecol Perspect Sci Soc 22:99–103. https://doi.org/10.14512/gaia.22.2.8

Nesshöver C, Vandewalle M, Wittmer H et al (2016) The network of knowledge approach: improving the science and society dialogue on biodiversity and ecosystem services in Europe. Biodivers Conserv 25:1215–1233. https://doi.org/10.1007/s10531-016-1127-5

Noss RF, Fleishman E, Dellasala DA et al (2009) Priorities for improving the scientific foundation of conservation policy in North America. Conserv Biol 23:825–833. https://doi.org/10.1111/j.1523-1739.2009.01282.x

Opdam PFM, Broekmeyer MEA, Kistenkas FH (2009) Identifying uncertainties in judging the significance of human impacts on Natura 2000 sites. Environ Sci Pol 12(7):912–921. https://doi.org/10.1016/j.envsci.2009.04.006

Paloniemi R, Apostolopoulou E, Primmer E et al (2012) Biodiversity conservation across scales: lessons from a science–policy dialogue. Nat Conserv 2:7–19

Plant R, Ryan P (2013) Ecosystem services as a practicable concept for natural resource management: some lessons from Australia. Int J Biodivers Sci Ecosyst Serv Manag 9:44–53. https://doi.org/10.1080/21513732.2012.737372

Pullin AS, Stewart GB (2006) Guidelines for systematic review in conservation and environmental management guidelines for systematic review in conservation and. Environ Manag 20:1647–1656. https://doi.org/10.1111/j.1523-1739.2006.00485.x

Pullin AS, Báldi A, Can OE et al (2009) Conservation focus on Europe: major conservation policy issues that need to be informed by conservation science. Conserv Biol 23:818–824. https://doi.org/10.1111/j.1523-1739.2009.01283.x

Raina RS, Dey D (2015) The valuation conundrum: biodiversity and science-policy interface in India's livestock sector. Econ Pol Wkly 51:7–8

Rodela R, Reinecke S, Bregt A et al (2015) Challenges to and opportunities for biodiversity science-policy interfaces. Environ Sci Pol 54:483–486. https://doi.org/10.1016/j.envsci.2015.08.010

Rodriguez MDL, Castro AJ, Jorreto S (2015) Science–policy interface for addressing environmental problems in arid Spain. Environ Sci Policy 50:1–14. https://doi.org/10.1016/j.envsci.2015.01.013

Roqueplo P (1995) Scientific expertise among political powers, administrations and public opinion. Sci Public Policy 22:175–182. https://doi.org/10.1093/spp/22.3.175

Ruckelshaus M, McKenzie E, Tallis H et al (2015) Notes from the field: lessons learned from using ecosystem service approaches to inform real-world decisions. Ecol Econ 115:11–21. https://doi.org/10.1016/j.ecolecon.2013.07.009

Sanguinetti J, Buria L, Malmierca L et al (2014) Invasive alien species management in Patagonia, Argentina: prioritization, achievements and science-policy integration challenges identified by the National Parks Administration. Ecol Austral 24:183–192

Santos MB, Pierce GJ (2015) Marine mammals and good environmental status: science, policy and society; challenges and opportunities. Hydrobiologia 750:13–41. https://doi.org/10.1007/s10750-014-2164-2

Sarkki S, Niemela J, Tinch R et al (2013) Balancing credibility, relevance and legitimacy: a critical assessment of trade-offs in science-policy interfaces. Sci Public Policy 41:194–206

Sarkki S, Tinch R, Niemelä J et al (2015) Adding "iterativity" to the credibility, relevance, legitimacy: a novel scheme to highlight dynamic aspects of science-policy interfaces. Environ Sci Pol 54:505–512. https://doi.org/10.1016/j.envsci.2015.02.016

Schewenius M, McPhearson T, Elmqvist T (2014) Opportunities for increasing resilience and sustainability of urban social–ecological systems: insights from the URBES and the Cities and Biodiversity Outlook Projects. Ambio 43:434–444. https://doi.org/10.1007/s13280-014-0505-z

Seddon N, Mace GM, Naeem S et al (2016) Biodiversity in the anthropocene: prospects and policy. Proc R Soc B Biol Sci 283:20162094. https://doi.org/10.1098/rspb.2016.2094

Sommerwerk N, Bloesch J, Paunovi M et al (2010a) Managing the worlds most international river: the Danube River Basin. Mar Freshw Res 61(7):736–748. https://doi.org/10.1071/MF09229

Sommerwerk N, Bloesch J, Paunović M et al (2010b) Managing the world's most international river: the Danube River Basin. Mar Freshw Res 61:736. https://doi.org/10.1071/MF09229

Spranger T, Bull K, Clair TA, Johansson M (2014) Implications of current knowledge on nitrogen deposition and impacts for policy, management and capacity building needs: CLRTAP. In: Nitrogen deposition, critical loads and biodiversity. Springer, Dordrecht, pp 425–434

Srebotnjak T (2007) The role of environmental statisticians in environmental policy: the case of performance measurement. Environ Sci Pol 10(5):405–418. https://doi.org/10.1016/j.envsci.2007.02.002

Stephenson PJ, Bowles-Newark N, Regan E et al (2015) Unblocking the flow of biodiversity data for decision-making in Africa. Biol Conserv 213:335–340. https://doi.org/10.1016/j.biocon.2016.09.003

Sutherland WJ, Pullin AS, Dolman PM, Knight TM (2004) The need for evidence-based conservation. Trends Ecol Evol 19:4–7. https://doi.org/10.1016/j.tree.2004.03.018

Sutherland WJ, Gardner T, Bogich TL et al (2014) Solution scanning as a key policy tool: identifying management interventions to help maintain and enhanc regulating ecosystem services. Ecol Soc 19(2):3

Thomas RJ, Akhtar-Schuster M, Stringer LC et al (2012) Fertile ground? Options for a science-policy platform for land. Environ Sci Pol 16:122–135

Tinch R, Balian E, Carss D et al (2016) Science-policy interfaces for biodiversity: dynamic learning environments for successful impact. Biodivers Conserv 27(7):1679–1702. https://doi.org/10.1007/s10531-016-1155-1

Tremblay M, Vandewalle M, Wittmer H (2016) Ethical challenges at the science-policy interface: an ethical risk assessment and proposition of an ethical infrastructure. Biodivers Conserv 25(7):1253–1267. https://doi.org/10.1007/s10531-016-1123-9

Tzankova Z (2017) The science and politics of ecological risk : bioinvasions policies in the US and Australia. Environ Pol 18(3):333–350. https://doi.org/10.1080/09644010902823725

van den Hove S (2007) A rationale for science–policy interfaces. Futures 39:807–826. https://doi.org/10.1016/j.futures.2006.12.004

van Eeten MJG (1999) "Dialogues of the deaf" on science in policy controversies. Sci Public Policy 26:185–192. https://doi.org/10.3152/147154399781782491

Van Haastrecht EK, Toonen HM (2011) Science-policy interactions in MPA site selection in the Dutch part of the North Sea. Environ. Manage 47(4):656–670

Vohland K, Mlambo MC, Horta LD et al (2011) How to ensure a credible and efficient IPBES? Environ Sci Pol 14(8):1188–1194. https://doi.org/10.1016/j.envsci.2011.08.005

Walther BA, Boëte C, Binot A et al (2016) Biodiversity and health: lessons and recommendations from an interdisciplinary conference to advise southeast Asian research, society and policy. Infect Genet Evol 40:29–46. https://doi.org/10.1016/j.meegid.2016.02.003

Watson RT (2012) The science–policy interface: the role of scientific assessments—UK National Ecosystem Assessment. Proc R Soc A 468:3265–3281. https://doi.org/10.1098/rspa.2012.0163

Weingart P (1999) Scientific expertise and political accountability: paradoxes of science in politics. Sci Public Policy 26:151–161. https://doi.org/10.3152/147154399781782437

Young JC, Watt AD, van den Hove S, the SPIRAL project team (2013a) Effective interfaces between science, policy and society: the SPIRAL project handbook

Young JC, Watt AD, van den Hove S, the SPIRAL project team (2013b) The SPIRAL synthesis report: a resource book on science-policy interfaces

Young JC, Waylen KA, Sarkki S et al (2014) Improving the science-policy dialogue to meet the challenges of biodiversity conservation: having conversations rather than talking at one-another. Biodivers Conserv 23:387–404. https://doi.org/10.1007/s10531-013-0607-0

Chapter 9
Synthesis: Managing Socio-ecological Production Landscapes and Seascapes for Sustainable Communities in Asia

Osamu Saito, Suneetha M Subramanian, Shizuka Hashimoto, and Kazuhiko Takeuchi

Abstract While Chaps. 2–5 covered specific case studies of landscapes and seascapes in Japan (Chaps. 2–4) and Bangladesh (Chap. 5), Chaps. 6–8 consisted of a series of review articles on sustainable management approaches relating to land/seascapes that explored lessons learned from assessing resilience in socio-ecological production landscapes and seascapes (SEPLS) (Chap. 6), solutions for sustainable management of SEPLS in Asia (Chap. 7), and the effectiveness of biodiversity science–policy interfaces (SPIs) from local to global scales (Chap. 8). These chapters are summarized here according to their objectives, materials/study sites, methods/tools, spatial scales, and key actors. Then, the implications for the United Nations Convention on Biological Diversity (CBD) Post-2020 Global Biodiversity Framework are discussed using key leverage points of transformations toward sustainability identified by the Intergovernmental Science-Policy Platform on Biodiversity and Ecosystem Services (IPBES) Global Assessment: (1) visions of a good life; (2) total consumption and waste; (3) values and action; (4) inequalities;

O. Saito (✉)
United Nations University Institute for the Advanced Study of Sustainability (UNU-IAS), Shibuya, Tokyo, Japan

Institute for Global Environmental Strategies (IGES), Hayama, Kanagawa, Japan

Institute for Future Initiatives (IFI), The University of Tokyo, Bunkyo, Tokyo, Japan
e-mail: saito@unu.edu

S. M. Subramanian
United Nations University International Institute for Global Health (UNU-IIGH), Cheras, Kuala Lumpur, Malaysia

S. Hashimoto
Graduate School of Agriculture and Life Sciences, The University of Tokyo, Bunkyo, Tokyo, Japan

K. Takeuchi
Institute for Global Environmental Strategies (IGES), Hayama, Kanagawa, Japan

Institute for Future Initiatives (IFI), The University of Tokyo, Bunkyo, Tokyo, Japan

© The Author(s) 2020
O. Saito et al. (eds.), *Managing Socio-ecological Production Landscapes and Seascapes for Sustainable Communities in Asia*, Science for Sustainable Societies, https://doi.org/10.1007/978-981-15-1133-2_9

(5) justice and inclusion in conservation; (6) externalities and telecoupling; (7) technology, innovation, and investment; and (8) education and knowledge generation and sharing.

Keywords Socio-ecological production landscapes and seascapes · Ecosystem services · Visualization · Mapping · Stakeholder analysis · Science–policy interface

9.1 Summary of the Book

Broadly, this book highlights various approaches to achieving the sustainable use of resources and development for socio-ecological production landscapes and seascapes (SEPLS) from local to global scales. While Chaps. 2–5 covered specific case studies at landscapes and seascapes in Japan (Chaps. 2–4) and Bangladesh (Chap. 5), Chaps. 6–8 consisted of a series of review articles that explored lessons learned from assessing resilience in SEPLS (Chap. 6), solutions for sustainable management of SEPLS in Asia (Chap. 7), and the effectiveness of biodiversity science–policy interfaces (SPIs) from local to global scales (Chap. 8). These chapters are summarized in Table 9.1 according to their objectives, materials/study sites, methods/tools, spatial scales, and key actors.

Focusing on the Sekisei Lagoon, Okinawa Prefecture, at the southeastern tip of the Japanese archipelago, Chap. 2 examined the inter-relationships between the sectoral policy interventions by various marine-related ministries and the entire structure of the integrated ocean policy. This study developed the SES schematic, which summarized and visualized the main ecosystem structures, functions, use types, and stakeholders relating to the lagoon. This SES schematic can be used as a boundary object to facilitate knowledge exchange between various stakeholders, including policy makers, practitioners, and researchers, to share a common understanding of the current situation, and to co-create policy interventions for sustainable uses of not only the Sekisei Lagoon but also other types of ecosystem or natural capital.

Chapter 3 focused on quantifying the willingness of tourists to participate in invasive carp removal in nature-based tourism in Amami Oshima, Japan. The study found that most tourists would avoid participating in carp removal activities as a tour option without any financial discounts but that over one third of tourists were willing to work for carp removal based on their own motivations. This result suggests that tourists could play an important role in invasive alien species management.

Using the example of the city of Toyama in Japan, Chap. 4 focused on a participatory approach of backcasting scenario-making to identify ways of bringing together various perspectives for sustainable urban planning. The chapter concluded that, when governed in certain ways, citizen participatory approaches can realize a fairly good balance between diverged processes and converged outcomes of backcasting scenario-making on the issue of urban sustainability transitions.

Chapter 5 highlighted how local institutions and traditional knowledge can be incorporated when addressing sustainable use and the conservation of biodiversity, focusing on experiences from the Sundarbans area in Bangladesh. Following MEB approaches, the chapter concluded that human sociality-based conservation

Table 9.1 Objectives, materials, methods/tools, spatial scales, and key actors in Chaps. 2–8

Chapter	Objectives	Materials/study sites	Methods/tools	Spatial scale	Key actors
Chapter 2	Mapping the policy interventions on marine social-ecological systems	A case study of the Sekisei Lagoon, Southwest Japan	The Social-Ecological Systems (SES) schematic: an integrated diagram of the inter-relationships between the main ecosystem structures, ecosystem functions, ecosystem use types, and stakeholders	Local scale	Includes central and local governments, fishermen, fish processers, marine cultural services users, and marine energy/resource developers
Chapter 3	Evaluating tourist opinions concerning participating in invasive carp removal in nature-based tourism	An experimental survey in Amami Oshima, Japan (343 questionnaires returned by mail)	A choice experiment on canoe tours to evaluate tourists' willingness to pay for tour options as a means of promoting canoe tours	Island scale	Tourists, tour operators, and the local government
Chapter 4	Backcasting scenario-making for sustainable urban transformation	A case study in Toyama, Japan	Backcasting scenario-making via a citizen participatory workshop	Local (city) scale	Citizen participants
Chapter 5	Proposing actions and policy alternatives to reverse the process of degradation and to move toward transformative harmonious human–nature interactions	A case study of the Sundarbans in Bangladesh	A conceptual framework of SEPLS, human sociality, and sustainability Multiple evidence-based (MEB) approaches	A transboundary mangrove ecosystem on the great delta of the Ganges	Indigenous peoples and local communities (IPLCs), property right owners, governments, factories, and shrimp cultivators

(continued)

Table 9.1 (continued)

Chapter	Objectives	Materials/study sites	Methods/tools	Spatial scale	Key actors
Chapter 6	Lessons from "Indicators of Resilience in SEPLS"	A total of 34 landscapes and seascapes around the world	The indicators of resilience in SEPLS (20 resilience indicators)	Local (community) scale	Local community, indigenous and local knowledge holders, local governments, and policy makers
Chapter 7	Reviewing place-based solutions for conservation and the restoration of social-ecological production landscapes and seascapes in Asia	A total of 88 case studies from the International Partnership for the Satoyama Initiative (IPSI) in the South, East, and Southeast Asian regions	A societal-based solution scanning approach	From local to national scales	Academia, public sector, community, and non-governmental (civic and private sectors)
Chapter 8	Reviewing the effectiveness of biodiversity SPIs	A total of 96 SPI studies worldwide	Systematic literature review Key features of SPIs	From local to global scales	Government and academia as facilitators of SPIs

practices positively impact resilient indicators and help achieve the Aichi Biodiversity Targets.

Chapter 6 examined applying the resilience assessment process using an indicator-based approach at 34 sites (communities) of SEPLS in different regions of the world. The measurement criteria are defined by individual communities, and therefore, the outcomes are specific to those communities when understanding the multiple aspects of resilience and changes over time and identifying important issues for improving the resilience of a community. The most prominent benefit found when using the indicators is their value as a convening tool, bringing together multiple stakeholders in a landscape or seascape.

Chapter 7 identified various categories of place-based solutions for the sustainable management of SEPLS based on the experiences of partners from the South, East, and Southeast Asian countries of the International Partnership for Satoyama Initiative (IPSI). Sharing knowledge of various place-based solution types in different social-ecological contexts helps provide more purposeful and deliberate designs of SEPLS with multiple benefits.

Chapter 8 reviewed the effectiveness of biodiversity SPIs by examining the different features of effective SPIs, including capacity building, trust building, adaptability, and continuity. The chapter concluded that effective, interdisciplinary SPIs and timely and relevant inputs for policymakers are required to ensure more dynamic, iterative, and collaborative interactions between policymakers and other actors.

9.2 Implications for Transformative Changes toward Sustainability

The IPBES Global Assessment (IPBES 2019) stressed that "goals for conserving and sustainably using nature and achieving sustainability cannot be met by current trajectories, and goals for 2030 and beyond may only be achieved through transformative changes across economic, social, political and technological factors." It is necessary for us to conserve, restore, and use nature sustainably while simultaneously meeting other global societal goals through extensive efforts that foster transformative change. Transformations toward sustainability can be triggered by following key leverage points: (1) visions of a good life; (2) total consumption and waste; (3) values and action; (4) inequalities; (5) justice and inclusion in conservation; (6) externalities and telecoupling; (7) technology, innovation, and investment; and (8) education and knowledge generation and sharing (IPBES 2019). Regariding these leverage points, Table 9.2 summarizes relevant approaches and insights linked to these leverage points as highlighted by the different experiences captured in this book. The Convention on Biological Diversity (CBD) has also been advocating the need to have a systems approach to address conservation and human well-being

Table 9.2 Leverage points and relevant approaches/insights provided by the book

Leverage points	Relevant chapters	Relevant approaches and insights from the chapters
(1) Visions of a good life	Chapters 4 and 5	– Participatory approach of backcasting scenario-making (Chap. 4) – MEB approaches (Chap. 5) – Importance of IPLCs in the conservation and management process (Chap. 5)
(2) Total consumption and waste	Chapters 2, 3, and 6	– SES schematics can holistically summarize the main ecosystem structures, functions, use types, and stakeholders (Chap. 2) – Perception of nature-based tourism for the removal of invasive species (Chap. 3)
(3) Values and action		– The resilience indicator toolkit approach helps communities to define their aspirations, take stock of their ecosystem integrity and resources, and identify strategies to achieve goals under indicators that are defined by the community and suited to their context (Chap. 6)
(4) Inequalities	Chapters 6, 7, and 8	– The resilience assessment includes indicators for governance and social equity such as social equity (including gender equity) (Chap. 6)
(5) Justice and inclusion in conservation		– The inclusion of community-based mapping exercises into resilience assessment workshops (Chap. 6) – Inclusion is one of the subcategories under institutional and governance solutions (Chap. 7) – Credibility, relevance, and legitimacy of SPIs (Chap. 8) – Inclusion of policy perspectives into research projects (Chap. 8)
(6) Externalities and telecoupling	Chapter 2	– SES schematics can capture externalities and telecoupling by covering a wide range of users beyond the target landscapes and seascapes including the transport and energy/resource development sectors (Chap. 2)
(7) Technology, innovation, and investment	Chapters 5 and 7	– Innovation and diversification of livelihood patterns including innovative techniques in agriculture (Chap. 5) – Technological solutions are one of the key solution types to reduce the harmful impacts of various drivers of ecosystem change, as well as underinvestment in the development and diffusion of technologies, and could increase the efficiency of resource or ecosystem use (Chap. 7)
(8) Education and knowledge generation and sharing	Chapters 2 and 7	– SES schematics can visually facilitate role sharing and knowledge sharing between different relevant stakeholders across scales (Chap. 2) – Knowledge and cognitive solutions are another key solution type to address insufficient knowledge or the poor use of existing knowledge concerning ecosystem services and addresses information gaps and incorporates other forms of knowledge and information (Chap. 7)

needs and has been emphasizing the need to embark on transformative change and manage transitions toward sustainable pathways (CBD 2017, 2018).

9.3   Afterword: Future Research Directions

This book presents contemporary experiences and analyses of community-based approaches to the sustainable resource management of SEPLS primarily based on experiences in Asia. The different cases highlight several pertinent issues regarding land/seascape approaches. First, empirical evidence illustrating the relevance of landscape approaches to the conservation of natural resources, contributions to economies, and sustainable livelihoods is compelling. The landscape approach is by nature an integrated approach that cuts across sectoral divisions and various policy priorities (e.g., environment, rural development, water management, health, and food security) and has a systemic focus on both the ecological and social dimensions within the land/seascape.

This implies that the interconnectedness of natural and human systems is highly entrenched in such areas and that the utilization and management of resources, even if driven by contextual priorities, have certain broad similarities, including those related to maintaining the multifunctionality of the landscape and ensuring a diversity of resources, a diversity of income sources based on primary production and services, and endogenous approaches that integrate traditional and modern practices and knowledge to ensure more sustainable outcomes. However, the experiences related in this book also indicate that sustaining such an ideal and idyllic scenario is fraught with various challenges ranging from policy drivers, changing priorities of the local population, demographic changes, the impact of distant market forces, the erosion of traditional practices, the homogenization of cropping practices, and changes in land use.

Addressing these challenges requires a comprehensive approach beginning with a clear understanding of changes in the natural resources and in the various drivers of change and of the implications for a good quality of life for the population. Possible solutions and pathways for the development that are participatory and inclusive in nature and ensuring a good alignment between macro policy goals and landscape level priorities need to be identified and implemented. New solutions may require the creation of flexible legal frameworks that protects the interests of, and reduces political constraints for, collaborative efforts in land/seascapes (Plieninger et al. 2018). This also implies an enhanced mandate for future research priorities focused on integrated approaches to landscape management to build inventories on the management, natural state, and drivers of change; to develop methodologies that further high fidelity scenarios developed using participatory approaches involving stakeholders on the ground; and to ensure that actions are taken at multiple scales, including local, regional, and beyond, and are aligned with new conceptual and policy concepts related to nature's contributions to people and sustainable development goals (Saito and Ichikawa 2014; Saito 2017). A global IPBES assessment identified similar gaps in knowledge, research, and resources (IPBES 2019). It is clear from the case studies in this book that it is possible to compare various policy outcomes from real-world experiences. Such experiences also highlight the utility of incorporating other ways of knowing, including data and trends of natural resources as observed by local communities, well-being parameters, and related drivers of change that can enhance existing knowledge of these subjects.

The effectiveness of any policy is reflected in how it is adopted and deployed by the people who are considered the most proximate stakeholders, whether in terms of resource proximity and/or impacts of outcomes. To ensure adoption, it is important that policies are sensitive to the priorities and challenges of such stakeholders. The chapters in this book provide a snapshot of possible approaches to streamline local and mainstream socio-ecological goals. We hope that it will serve to foster more creative thinking and support toward the revitalization of dynamic socio-ecological systems, enabling locally led conservation actions and broad-based development across different regions of the world.

Acknowledgments This book was funded by the EnvironmentvResearch and Technology Development Fund (S-15 "Predicting and Assessing Natural Capital and Ecosystem Services" (PANCES), Ministry of the Environment, Japan). We also acknowledge various contributions by International Partnership for the Satoama Initiative (IPSI).

References

CBD (2017) Discussion note on toward a transformative change for biodiversity based on systems transition. https://www.cbd.int/cooperation/bogis/S111.pdf

CBD (2018) Press release. https://www.cbd.int/doc/press/2018/pr-2018-07-18-sbstta22-sbi2-en.pdf

IPBES (2019) Summary for policymakers of the global assessment report on biodiversity and ecosystem services of the Intergovernmental Science-Policy Platform on Biodiversity and Ecosystem Services, IPBES/7/10/Add.1. https://www.ipbes.net/system/tdf/ipbes_7_10_add-1-_advance_0.pdf?file=1&type=node&id=35245

Plieninger T, Kohsaka R, Bieling C, Hashimoto S, Kamiyama C, Kizos T, Penker M, Kieninger P, Shaw BJ, Sioen GB, Yoshida Y, Saito O (2018) Fostering biocultural diversity in landscapes through place-based food networks: a "solution scan" of European and Japanese models. Sustain Sci 13:1–15. https://doi.org/10.1007/s11625-017-0455-z

Saito O (2017) Future science-policy agendas and partnerships for building a sustainable society in harmony with nature. Sustain Sci 12:895–899. https://doi.org/10.1007/s11625-017-0475-8

Saito O, Ichikawa K (2014) Socio-ecological systems in paddy-dominated landscapes in Asian Monsoon. In: Nishikawa U, Miyashita T (eds) Social-ecological restoration in paddy-dominated landscapes. Springer, New York, pp 17–37